C-570 CAREER EXAMINATION SERIES

This is your PASSBOOK for...

Painter

Test Preparation Study Guide
Questions & Answers

NLC®
NATIONAL LEARNING CORPORATION®

COPYRIGHT NOTICE

National Learning Corporation

212 Michael Drive, Syosset, NY 11791
(516) 921-8888 • www.passbooks.com
E-mail: info@passbooks.com

PUBLISHED IN THE UNITED STATES OF AMERICA

PASSBOOK® SERIES

THE *PASSBOOK® SERIES* has been created to prepare applicants and candidates for the ultimate academic battlefield – the examination room.

At some time in our lives, each and every one of us may be required to take an examination – for validation, matriculation, admission, qualification, registration, certification, or licensure.

Based on the assumption that every applicant or candidate has met the basic formal educational standards, has taken the required number of courses, and read the necessary texts, the *PASSBOOK® SERIES* furnishes the one special preparation which may assure passing with confidence, instead of failing with insecurity. Examination questions – together with answers – are furnished as the basic vehicle for study so that the mysteries of the examination and its compounding difficulties may be eliminated or diminished by a sure method.

This book is meant to help you pass your examination provided that you qualify and are serious in your objective.

The entire field is reviewed through the huge store of content information which is succinctly presented through a provocative and challenging approach – the question-and-answer method.

A climate of success is established by furnishing the correct answers at the end of each test.

You soon learn to recognize types of questions, forms of questions, and patterns of questioning. You may even begin to anticipate expected outcomes.

You perceive that many questions are repeated or adapted so that you can gain acute insights, which may enable you to score many sure points.

You learn how to confront new questions, or types of questions, and to attack them confidently and work out the correct answers.

You note objectives and emphases, and recognize pitfalls and dangers, so that you may make positive educational adjustments.

Moreover, you are kept fully informed in relation to new concepts, methods, practices, and directions in the field.

You discover that you are actually taking the examination all the time: you are preparing for the examination by "taking" an examination, not by reading extraneous and/or supererogatory textbooks.

In short, this PASSBOOK®, used directedly, should be an important factor in helping you to pass your test.

PAINTER

DUTIES

Painters, under supervision, do inside and outside patching and painting of a general nature, including all coats; prepare, fill and prime surfaces for painting; erect ladders; may rig lines and scaffolds; mix paint components and match colors; apply paint to surfaces with a brush or roller or spray-gun; apply plaster to surfaces; take proper care of all materials, tools and related equipment; keep work records as required; may operate a motor vehicle; and perform related work.

TESTS

The multiple-choice test may include questions on techniques and materials used in painting and preparing surfaces; proper use of tools and equipment; ability to follow written instructions and fill out forms and/or reports; safety practices applicable to the painting trade; maintenance of tools and equipment; and other related areas.

HOW TO TAKE A TEST

I. YOU MUST PASS AN EXAMINATION

A. WHAT EVERY CANDIDATE SHOULD KNOW

Examination applicants often ask us for help in preparing for the written test. What can I study in advance? What kinds of questions will be asked? How will the test be given? How will the papers be graded?

As an applicant for a civil service examination, you may be wondering about some of these things. Our purpose here is to suggest effective methods of advance study and to describe civil service examinations.

Your chances for success on this examination can be increased if you know how to prepare. Those "pre-examination jitters" can be reduced if you know what to expect. You can even experience an adventure in good citizenship if you know why civil service exams are given.

B. WHY ARE CIVIL SERVICE EXAMINATIONS GIVEN?

Civil service examinations are important to you in two ways. As a citizen, you want public jobs filled by employees who know how to do their work. As a job seeker, you want a fair chance to compete for that job on an equal footing with other candidates. The best-known means of accomplishing this two-fold goal is the competitive examination.

Exams are widely publicized throughout the nation. They may be administered for jobs in federal, state, city, municipal, town or village governments or agencies.

Any citizen may apply, with some limitations, such as the age or residence of applicants. Your experience and education may be reviewed to see whether you meet the requirements for the particular examination. When these requirements exist, they are reasonable and applied consistently to all applicants. Thus, a competitive examination may cause you some uneasiness now, but it is your privilege and safeguard.

C. HOW ARE CIVIL SERVICE EXAMS DEVELOPED?

Examinations are carefully written by trained technicians who are specialists in the field known as "psychological measurement," in consultation with recognized authorities in the field of work that the test will cover. These experts recommend the subject matter areas or skills to be tested; only those knowledges or skills important to your success on the job are included. The most reliable books and source materials available are used as references. Together, the experts and technicians judge the difficulty level of the questions.

Test technicians know how to phrase questions so that the problem is clearly stated. Their ethics do not permit "trick" or "catch" questions. Questions may have been tried out on sample groups, or subjected to statistical analysis, to determine their usefulness.

Written tests are often used in combination with performance tests, ratings of training and experience, and oral interviews. All of these measures combine to form the best-known means of finding the right person for the right job.

II. HOW TO PASS THE WRITTEN TEST

A. NATURE OF THE EXAMINATION

To prepare intelligently for civil service examinations, you should know how they differ from school examinations you have taken. In school you were assigned certain definite pages to read or subjects to cover. The examination questions were quite detailed and usually emphasized memory. Civil service exams, on the other hand, try to discover your present ability to perform the duties of a position, plus your potentiality to learn these duties. In other words, a civil service exam attempts to predict how successful you will be. Questions cover such a broad area that they cannot be as minute and detailed as school exam questions.

In the public service similar kinds of work, or positions, are grouped together in one "class." This process is known as *position-classification*. All the positions in a class are paid according to the salary range for that class. One class title covers all of these positions, and they are all tested by the same examination.

B. FOUR BASIC STEPS

1) Study the announcement

How, then, can you know what subjects to study? Our best answer is: "Learn as much as possible about the class of positions for which you've applied." The exam will test the knowledge, skills and abilities needed to do the work.

Your most valuable source of information about the position you want is the official exam announcement. This announcement lists the training and experience qualifications. Check these standards and apply only if you come reasonably close to meeting them.

The brief description of the position in the examination announcement offers some clues to the subjects which will be tested. Think about the job itself. Review the duties in your mind. Can you perform them, or are there some in which you are rusty? Fill in the blank spots in your preparation.

Many jurisdictions preview the written test in the exam announcement by including a section called "Knowledge and Abilities Required," "Scope of the Examination," or some similar heading. Here you will find out specifically what fields will be tested.

2) Review your own background

Once you learn in general what the position is all about, and what you need to know to do the work, ask yourself which subjects you already know fairly well and which need improvement. You may wonder whether to concentrate on improving your strong areas or on building some background in your fields of weakness. When the announcement has specified "some knowledge" or "considerable knowledge," or has used adjectives like "beginning principles of..." or "advanced ... methods," you can get a clue as to the number and difficulty of questions to be asked in any given field. More questions, and hence broader coverage, would be included for those subjects which are more important in the work. Now weigh your strengths and weaknesses against the job requirements and prepare accordingly.

3) Determine the level of the position

Another way to tell how intensively you should prepare is to understand the level of the job for which you are applying. Is it the entering level? In other words, is this the position in which beginners in a field of work are hired? Or is it an intermediate or advanced level? Sometimes this is indicated by such words as "Junior" or "Senior" in the class title. Other jurisdictions use Roman numerals to designate the level – Clerk I, Clerk II, for example. The word "Supervisor" sometimes appears in the title. If the level is not indicated by the title, check the description of duties. Will you be working under very close supervision, or will you have responsibility for independent decisions in this work?

4) Choose appropriate study materials

Now that you know the subjects to be examined and the relative amount of each subject to be covered, you can choose suitable study materials. For beginning level jobs, or even advanced ones, if you have a pronounced weakness in some aspect of your training, read a modern, standard textbook in that field. Be sure it is up to date and has general coverage. Such books are normally available at your library, and the librarian will be glad to help you locate one. For entry-level positions, questions of appropriate difficulty are chosen – neither highly advanced questions, nor those too simple. Such questions require careful thought but not advanced training.

If the position for which you are applying is technical or advanced, you will read more advanced, specialized material. If you are already familiar with the basic principles of your field, elementary textbooks would waste your time. Concentrate on advanced textbooks and technical periodicals. Think through the concepts and review difficult problems in your field.

These are all general sources. You can get more ideas on your own initiative, following these leads. For example, training manuals and publications of the government agency which employs workers in your field can be useful, particularly for technical and professional positions. A letter or visit to the government department involved may result in more specific study suggestions, and certainly will provide you with a more definite idea of the exact nature of the position you are seeking.

III. KINDS OF TESTS

Tests are used for purposes other than measuring knowledge and ability to perform specified duties. For some positions, it is equally important to test ability to make adjustments to new situations or to profit from training. In others, basic mental abilities not dependent on information are essential. Questions which test these things may not appear as pertinent to the duties of the position as those which test for knowledge and information. Yet they are often highly important parts of a fair examination. For very general questions, it is almost impossible to help you direct your study efforts. What we can do is to point out some of the more common of these general abilities needed in public service positions and describe some typical questions.

1) General information

Broad, general information has been found useful for predicting job success in some kinds of work. This is tested in a variety of ways, from vocabulary lists to questions about current events. Basic background in some field of work, such as sociology or economics, may be sampled in a group of questions. Often these are principles which have become familiar to most persons through exposure rather than through formal training. It is difficult to advise you how to study for these questions; being alert to the world around you is our best suggestion.

2) Verbal ability

An example of an ability needed in many positions is verbal or language ability. Verbal ability is, in brief, the ability to use and understand words. Vocabulary and grammar tests are typical measures of this ability. Reading comprehension or paragraph interpretation questions are common in many kinds of civil service tests. You are given a paragraph of written material and asked to find its central meaning.

3) Numerical ability

Number skills can be tested by the familiar arithmetic problem, by checking paired lists of numbers to see which are alike and which are different, or by interpreting charts and graphs. In the latter test, a graph may be printed in the test booklet which you are asked to use as the basis for answering questions.

4) Observation

A popular test for law-enforcement positions is the observation test. A picture is shown to you for several minutes, then taken away. Questions about the picture test your ability to observe both details and larger elements.

5) Following directions

In many positions in the public service, the employee must be able to carry out written instructions dependably and accurately. You may be given a chart with several columns, each column listing a variety of information. The questions require you to carry out directions involving the information given in the chart.

6) Skills and aptitudes

Performance tests effectively measure some manual skills and aptitudes. When the skill is one in which you are trained, such as typing or shorthand, you can practice. These tests are often very much like those given in business school or high school courses. For many of the other skills and aptitudes, however, no short-time preparation can be made. Skills and abilities natural to you or that you have developed throughout your lifetime are being tested.

Many of the general questions just described provide all the data needed to answer the questions and ask you to use your reasoning ability to find the answers. Your best preparation for these tests, as well as for tests of facts and ideas, is to be at your physical and mental best. You, no doubt, have your own methods of getting into an exam-taking mood and keeping "in shape." The next section lists some ideas on this subject.

IV. KINDS OF QUESTIONS

Only rarely is the "essay" question, which you answer in narrative form, used in civil service tests. Civil service tests are usually of the short-answer type. Full instructions for answering these questions will be given to you at the examination. But in case this is your first experience with short-answer questions and separate answer sheets, here is what you need to know:

1) Multiple-choice Questions

Most popular of the short-answer questions is the "multiple choice" or "best answer" question. It can be used, for example, to test for factual knowledge, ability to solve problems or judgment in meeting situations found at work.

A multiple-choice question is normally one of three types—

- It can begin with an incomplete statement followed by several possible endings. You are to find the one ending which *best* completes the statement, although some of the others may not be entirely wrong.
- It can also be a complete statement in the form of a question which is answered by choosing one of the statements listed.

- It can be in the form of a problem – again you select the best answer.

Here is an example of a multiple-choice question with a discussion which should give you some clues as to the method for choosing the right answer:

When an employee has a complaint about his assignment, the action which will *best* help him overcome his difficulty is to

A. discuss his difficulty with his coworkers
B. take the problem to the head of the organization
C. take the problem to the person who gave him the assignment
D. say nothing to anyone about his complaint

In answering this question, you should study each of the choices to find which is best. Consider choice "A" – Certainly an employee may discuss his complaint with fellow employees, but no change or improvement can result, and the complaint remains unresolved. Choice "B" is a poor choice since the head of the organization probably does not know what assignment you have been given, and taking your problem to him is known as "going over the head" of the supervisor. The supervisor, or person who made the assignment, is the person who can clarify it or correct any injustice. Choice "C" is, therefore, correct. To say nothing, as in choice "D," is unwise. Supervisors have and interest in knowing the problems employees are facing, and the employee is seeking a solution to his problem.

2) True/False Questions

The "true/false" or "right/wrong" form of question is sometimes used. Here a complete statement is given. Your job is to decide whether the statement is right or wrong.

SAMPLE: A roaming cell-phone call to a nearby city costs less than a non-roaming call to a distant city.

This statement is wrong, or false, since roaming calls are more expensive.

This is not a complete list of all possible question forms, although most of the others are variations of these common types. You will always get complete directions for answering questions. Be sure you understand *how* to mark your answers – ask questions until you do.

V. RECORDING YOUR ANSWERS

Computer terminals are used more and more today for many different kinds of exams.

For an examination with very few applicants, you may be told to record your answers in the test booklet itself. Separate answer sheets are much more common. If this separate answer sheet is to be scored by machine – and this is often the case – it is highly important that you mark your answers correctly in order to get credit.

An electronic scoring machine is often used in civil service offices because of the speed with which papers can be scored. Machine-scored answer sheets must be marked with a pencil, which will be given to you. This pencil has a high graphite content which responds to the electronic scoring machine. As a matter of fact, stray dots may register as answers, so do not let your pencil rest on the answer sheet while you are pondering the correct answer. Also, if your pencil lead breaks or is otherwise defective, ask for another.

Since the answer sheet will be dropped in a slot in the scoring machine, be careful not to bend the corners or get the paper crumpled.

The answer sheet normally has five vertical columns of numbers, with 30 numbers to a column. These numbers correspond to the question numbers in your test booklet. After each number, going across the page are four or five pairs of dotted lines. These short dotted lines have small letters or numbers above them. The first two pairs may also have a "T" or "F" above the letters. This indicates that the first two pairs only are to be used if the questions are of the true-false type. If the questions are multiple choice, disregard the "T" and "F" and pay attention only to the small letters or numbers.

Answer your questions in the manner of the sample that follows:

32. The largest city in the United States is
    A. Washington, D.C.
    B. New York City
    C. Chicago
    D. Detroit
    E. San Francisco

1) Choose the answer you think is best. (New York City is the largest, so "B" is correct.)
2) Find the row of dotted lines numbered the same as the question you are answering. (Find row number 32)
3) Find the pair of dotted lines corresponding to the answer. (Find the pair of lines under the mark "B.")
4) Make a solid black mark between the dotted lines.

VI. BEFORE THE TEST

Common sense will help you find procedures to follow to get ready for an examination. Too many of us, however, overlook these sensible measures. Indeed, nervousness and fatigue have been found to be the most serious reasons why applicants fail to do their best on civil service tests. Here is a list of reminders:

- Begin your preparation early – Don't wait until the last minute to go scurrying around for books and materials or to find out what the position is all about.
- Prepare continuously – An hour a night for a week is better than an all-night cram session. This has been definitely established. What is more, a night a week for a month will return better dividends than crowding your study into a shorter period of time.
- Locate the place of the exam – You have been sent a notice telling you when and where to report for the examination. If the location is in a different town or otherwise unfamiliar to you, it would be well to inquire the best route and learn something about the building.
- Relax the night before the test – Allow your mind to rest. Do not study at all that night. Plan some mild recreation or diversion; then go to bed early and get a good night's sleep.
- Get up early enough to make a leisurely trip to the place for the test – This way unforeseen events, traffic snarls, unfamiliar buildings, etc. will not upset you.
- Dress comfortably – A written test is not a fashion show. You will be known by number and not by name, so wear something comfortable.

- Leave excess paraphernalia at home – Shopping bags and odd bundles will get in your way. You need bring only the items mentioned in the official notice you received; usually everything you need is provided. Do not bring reference books to the exam. They will only confuse those last minutes and be taken away from you when in the test room.
- Arrive somewhat ahead of time – If because of transportation schedules you must get there very early, bring a newspaper or magazine to take your mind off yourself while waiting.
- Locate the examination room – When you have found the proper room, you will be directed to the seat or part of the room where you will sit. Sometimes you are given a sheet of instructions to read while you are waiting. Do not fill out any forms until you are told to do so; just read them and be prepared.
- Relax and prepare to listen to the instructions
- If you have any physical problem that may keep you from doing your best, be sure to tell the test administrator. If you are sick or in poor health, you really cannot do your best on the exam. You can come back and take the test some other time.

VII. AT THE TEST

The day of the test is here and you have the test booklet in your hand. The temptation to get going is very strong. Caution! There is more to success than knowing the right answers. You must know how to identify your papers and understand variations in the type of short-answer question used in this particular examination. Follow these suggestions for maximum results from your efforts:

1) Cooperate with the monitor

The test administrator has a duty to create a situation in which you can be as much at ease as possible. He will give instructions, tell you when to begin, check to see that you are marking your answer sheet correctly, and so on. He is not there to guard you, although he will see that your competitors do not take unfair advantage. He wants to help you do your best.

2) Listen to all instructions

Don't jump the gun! Wait until you understand all directions. In most civil service tests you get more time than you need to answer the questions. So don't be in a hurry. Read each word of instructions until you clearly understand the meaning. Study the examples, listen to all announcements and follow directions. Ask questions if you do not understand what to do.

3) Identify your papers

Civil service exams are usually identified by number only. You will be assigned a number; you must not put your name on your test papers. Be sure to copy your number correctly. Since more than one exam may be given, copy your exact examination title.

4) Plan your time

Unless you are told that a test is a "speed" or "rate of work" test, speed itself is usually not important. Time enough to answer all the questions will be provided, but this does not mean that you have all day. An overall time limit has been set. Divide the total time (in minutes) by the number of questions to determine the approximate time you have for each question.

5) Do not linger over difficult questions

If you come across a difficult question, mark it with a paper clip (useful to have along) and come back to it when you have been through the booklet. One caution if you do this – be sure to skip a number on your answer sheet as well. Check often to be sure that you have not lost your place and that you are marking in the row numbered the same as the question you are answering.

6) Read the questions

Be sure you know what the question asks! Many capable people are unsuccessful because they failed to *read* the questions correctly.

7) Answer all questions

Unless you have been instructed that a penalty will be deducted for incorrect answers, it is better to guess than to omit a question.

8) Speed tests

It is often better NOT to guess on speed tests. It has been found that on timed tests people are tempted to spend the last few seconds before time is called in marking answers at random – without even reading them – in the hope of picking up a few extra points. To discourage this practice, the instructions may warn you that your score will be "corrected" for guessing. That is, a penalty will be applied. The incorrect answers will be deducted from the correct ones, or some other penalty formula will be used.

9) Review your answers

If you finish before time is called, go back to the questions you guessed or omitted to give them further thought. Review other answers if you have time.

10) Return your test materials

If you are ready to leave before others have finished or time is called, take ALL your materials to the monitor and leave quietly. Never take any test material with you. The monitor can discover whose papers are not complete, and taking a test booklet may be grounds for disqualification.

VIII. EXAMINATION TECHNIQUES

1) Read the general instructions carefully. These are usually printed on the first page of the exam booklet. As a rule, these instructions refer to the timing of the examination; the fact that you should not start work until the signal and must stop work at a signal, etc. If there are any *special* instructions, such as a choice of questions to be answered, make sure that you note this instruction carefully.

2) When you are ready to start work on the examination, that is as soon as the signal has been given, read the instructions to each question booklet, underline any key words or phrases, such as *least, best, outline, describe* and the like. In this way you will tend to answer as requested rather than discover on reviewing your paper that you *listed without describing*, that you selected the *worst* choice rather than the *best* choice, etc.

3) If the examination is of the objective or multiple-choice type – that is, each question will also give a series of possible answers: A, B, C or D, and you are called upon to select the best answer and write the letter next to that answer on your answer paper – it is advisable to start answering each question in turn. There may be anywhere from 50 to 100 such questions in the three or four hours allotted and you can see how much time would be taken if you read through all the questions before beginning to answer any. Furthermore, if you come across a question or group of questions which you know would be difficult to answer, it would undoubtedly affect your handling of all the other questions.

4) If the examination is of the essay type and contains but a few questions, it is a moot point as to whether you should read all the questions before starting to answer any one. Of course, if you are given a choice – say five out of seven and the like – then it is essential to read all the questions so you can eliminate the two that are most difficult. If, however, you are asked to answer all the questions, there may be danger in trying to answer the easiest one first because you may find that you will spend too much time on it. The best technique is to answer the first question, then proceed to the second, etc.

5) Time your answers. Before the exam begins, write down the time it started, then add the time allowed for the examination and write down the time it must be completed, then divide the time available somewhat as follows:
   - If 3-1/2 hours are allowed, that would be 210 minutes. If you have 80 objective-type questions, that would be an average of 2-1/2 minutes per question. Allow yourself no more than 2 minutes per question, or a total of 160 minutes, which will permit about 50 minutes to review.
   - If for the time allotment of 210 minutes there are 7 essay questions to answer, that would average about 30 minutes a question. Give yourself only 25 minutes per question so that you have about 35 minutes to review.

6) The most important instruction is to *read each question* and make sure you know what is wanted. The second most important instruction is to *time yourself properly* so that you answer every question. The third most important instruction is to *answer every question.* Guess if you have to but include something for each question. Remember that you will receive no credit for a blank and will probably receive some credit if you write something in answer to an essay question. If you guess a letter – say "B" for a multiple-choice question – you may have guessed right. If you leave a blank as an answer to a multiple-choice question, the examiners may respect your feelings but it will not add a point to your score. Some exams may penalize you for wrong answers, so in such cases *only*, you may not want to guess unless you have some basis for your answer.

7) Suggestions
   a. Objective-type questions
      1. Examine the question booklet for proper sequence of pages and questions
      2. Read all instructions carefully
      3. Skip any question which seems too difficult; return to it after all other questions have been answered
      4. Apportion your time properly; do not spend too much time on any single question or group of questions

5. Note and underline key words – *all, most, fewest, least, best, worst, same, opposite*, etc.
6. Pay particular attention to negatives
7. Note unusual option, e.g., unduly long, short, complex, different or similar in content to the body of the question
8. Observe the use of "hedging" words – *probably, may, most likely*, etc.
9. Make sure that your answer is put next to the same number as the question
10. Do not second-guess unless you have good reason to believe the second answer is definitely more correct
11. Cross out original answer if you decide another answer is more accurate; do not erase until you are ready to hand your paper in
12. Answer all questions; guess unless instructed otherwise
13. Leave time for review

b. Essay questions

1. Read each question carefully
2. Determine exactly what is wanted. Underline key words or phrases.
3. Decide on outline or paragraph answer
4. Include many different points and elements unless asked to develop any one or two points or elements
5. Show impartiality by giving pros and cons unless directed to select one side only
6. Make and write down any assumptions you find necessary to answer the questions
7. Watch your English, grammar, punctuation and choice of words
8. Time your answers; don't crowd material

8) Answering the essay question

Most essay questions can be answered by framing the specific response around several key words or ideas. Here are a few such key words or ideas:

M's: manpower, materials, methods, money, management
P's: purpose, program, policy, plan, procedure, practice, problems, pitfalls, personnel, public relations

a. Six basic steps in handling problems:

1. Preliminary plan and background development
2. Collect information, data and facts
3. Analyze and interpret information, data and facts
4. Analyze and develop solutions as well as make recommendations
5. Prepare report and sell recommendations
6. Install recommendations and follow up effectiveness

b. Pitfalls to avoid

1. *Taking things for granted* – A statement of the situation does not necessarily imply that each of the elements is necessarily true; for example, a complaint may be invalid and biased so that all that can be taken for granted is that a complaint has been registered

2. *Considering only one side of a situation* – Wherever possible, indicate several alternatives and then point out the reasons you selected the best one
3. *Failing to indicate follow up* – Whenever your answer indicates action on your part, make certain that you will take proper follow-up action to see how successful your recommendations, procedures or actions turn out to be
4. *Taking too long in answering any single question* – Remember to time your answers properly

IX. AFTER THE TEST

Scoring procedures differ in detail among civil service jurisdictions although the general principles are the same. Whether the papers are hand-scored or graded by machine we have described, they are nearly always graded by number. That is, the person who marks the paper knows only the number – never the name – of the applicant. Not until all the papers have been graded will they be matched with names. If other tests, such as training and experience or oral interview ratings have been given, scores will be combined. Different parts of the examination usually have different weights. For example, the written test might count 60 percent of the final grade, and a rating of training and experience 40 percent. In many jurisdictions, veterans will have a certain number of points added to their grades.

After the final grade has been determined, the names are placed in grade order and an eligible list is established. There are various methods for resolving ties between those who get the same final grade – probably the most common is to place first the name of the person whose application was received first. Job offers are made from the eligible list in the order the names appear on it. You will be notified of your grade and your rank as soon as all these computations have been made. This will be done as rapidly as possible.

People who are found to meet the requirements in the announcement are called "eligibles." Their names are put on a list of eligible candidates. An eligible's chances of getting a job depend on how high he stands on this list and how fast agencies are filling jobs from the list.

When a job is to be filled from a list of eligibles, the agency asks for the names of people on the list of eligibles for that job. When the civil service commission receives this request, it sends to the agency the names of the three people highest on this list. Or, if the job to be filled has specialized requirements, the office sends the agency the names of the top three persons who meet these requirements from the general list.

The appointing officer makes a choice from among the three people whose names were sent to him. If the selected person accepts the appointment, the names of the others are put back on the list to be considered for future openings.

That is the rule in hiring from all kinds of eligible lists, whether they are for typist, carpenter, chemist, or something else. For every vacancy, the appointing officer has his choice of any one of the top three eligibles on the list. This explains why the person whose name is on top of the list sometimes does not get an appointment when some of the persons lower on the list do. If the appointing officer chooses the second or third eligible, the No. 1 eligible does not get a job at once, but stays on the list until he is appointed or the list is terminated.

X. HOW TO PASS THE INTERVIEW TEST

The examination for which you applied requires an oral interview test. You have already taken the written test and you are now being called for the interview test – the final part of the formal examination.

You may think that it is not possible to prepare for an interview test and that there are no procedures to follow during an interview. Our purpose is to point out some things you can do in advance that will help you and some good rules to follow and pitfalls to avoid while you are being interviewed.

What is an interview supposed to test?

The written examination is designed to test the technical knowledge and competence of the candidate; the oral is designed to evaluate intangible qualities, not readily measured otherwise, and to establish a list showing the relative fitness of each candidate – as measured against his competitors – for the position sought. Scoring is not on the basis of "right" and "wrong," but on a sliding scale of values ranging from "not passable" to "outstanding." As a matter of fact, it is possible to achieve a relatively low score without a single "incorrect" answer because of evident weakness in the qualities being measured.

Occasionally, an examination may consist entirely of an oral test – either an individual or a group oral. In such cases, information is sought concerning the technical knowledges and abilities of the candidate, since there has been no written examination for this purpose. More commonly, however, an oral test is used to supplement a written examination.

Who conducts interviews?

The composition of oral boards varies among different jurisdictions. In nearly all, a representative of the personnel department serves as chairman. One of the members of the board may be a representative of the department in which the candidate would work. In some cases, "outside experts" are used, and, frequently, a businessman or some other representative of the general public is asked to serve. Labor and management or other special groups may be represented. The aim is to secure the services of experts in the appropriate field.

However the board is composed, it is a good idea (and not at all improper or unethical) to ascertain in advance of the interview who the members are and what groups they represent. When you are introduced to them, you will have some idea of their backgrounds and interests, and at least you will not stutter and stammer over their names.

What should be done before the interview?

While knowledge about the board members is useful and takes some of the surprise element out of the interview, there is other preparation which is more substantive. It *is* possible to prepare for an oral interview – in several ways:

1) Keep a copy of your application and review it carefully before the interview

This may be the only document before the oral board, and the starting point of the interview. Know what education and experience you have listed there, and the sequence and dates of all of it. Sometimes the board will ask you to review the highlights of your experience for them; you should not have to hem and haw doing it.

2) Study the class specification and the examination announcement

Usually, the oral board has one or both of these to guide them. The qualities, characteristics or knowledges required by the position sought are stated in these documents. They offer valuable clues as to the nature of the oral interview. For example, if the job involves supervisory responsibilities, the announcement will usually indicate that knowledge of modern supervisory methods and the qualifications of the candidate as a supervisor will be tested. If so, you can expect such questions, frequently in the form of a hypothetical situation which you are expected to solve. NEVER go into an oral without knowledge of the duties and responsibilities of the job you seek.

3) Think through each qualification required

Try to visualize the kind of questions you would ask if you were a board member. How well could you answer them? Try especially to appraise your own knowledge and background in each area, *measured against the job sought*, and identify any areas in which you are weak. Be critical and realistic – do not flatter yourself.

4) Do some general reading in areas in which you feel you may be weak

For example, if the job involves supervision and your past experience has NOT, some general reading in supervisory methods and practices, particularly in the field of human relations, might be useful. Do NOT study agency procedures or detailed manuals. The oral board will be testing your understanding and capacity, not your memory.

5) Get a good night's sleep and watch your general health and mental attitude

You will want a clear head at the interview. Take care of a cold or any other minor ailment, and of course, no hangovers.

What should be done on the day of the interview?

Now comes the day of the interview itself. Give yourself plenty of time to get there. Plan to arrive somewhat ahead of the scheduled time, particularly if your appointment is in the fore part of the day. If a previous candidate fails to appear, the board might be ready for you a bit early. By early afternoon an oral board is almost invariably behind schedule if there are many candidates, and you may have to wait. Take along a book or magazine to read, or your application to review, but leave any extraneous material in the waiting room when you go in for your interview. In any event, relax and compose yourself.

The matter of dress is important. The board is forming impressions about you – from your experience, your manners, your attitude, and your appearance. Give your personal appearance careful attention. Dress your best, but not your flashiest. Choose conservative, appropriate clothing, and be sure it is immaculate. This is a business interview, and your appearance should indicate that you regard it as such. Besides, being well groomed and properly dressed will help boost your confidence.

Sooner or later, someone will call your name and escort you into the interview room. *This is it.* From here on you are on your own. It is too late for any more preparation. But remember, you asked for this opportunity to prove your fitness, and you are here because your request was granted.

What happens when you go in?

The usual sequence of events will be as follows: The clerk (who is often the board stenographer) will introduce you to the chairman of the oral board, who will introduce you to the other members of the board. Acknowledge the introductions before you sit down. Do not be surprised if you find a microphone facing you or a stenotypist sitting by. Oral interviews are usually recorded in the event of an appeal or other review.

Usually the chairman of the board will open the interview by reviewing the highlights of your education and work experience from your application – primarily for the benefit of the other members of the board, as well as to get the material into the record. Do not interrupt or comment unless there is an error or significant misinterpretation; if that is the case, do not hesitate. But do not quibble about insignificant matters. Also, he will usually ask you some question about your education, experience or your present job – partly to get you to start talking and to establish the interviewing "rapport." He may start the actual questioning, or turn it over to one of the other members. Frequently, each member undertakes the questioning on a particular area, one in which he is perhaps most competent, so you can expect each member to participate in the examination. Because time is limited, you may also expect some rather abrupt switches in the direction the questioning takes, so do not be upset by it. Normally, a board member will not pursue a single line of questioning unless he discovers a particular strength or weakness.

After each member has participated, the chairman will usually ask whether any member has any further questions, then will ask you if you have anything you wish to add. Unless you are expecting this question, it may floor you. Worse, it may start you off on an extended, extemporaneous speech. The board is not usually seeking more information. The question is principally to offer you a last opportunity to present further qualifications or to indicate that you have nothing to add. So, if you feel that a significant qualification or characteristic has been overlooked, it is proper to point it out in a sentence or so. Do not compliment the board on the thoroughness of their examination – they have been sketchy, and you know it. If you wish, merely say, "No thank you, I have nothing further to add." This is a point where you can "talk yourself out" of a good impression or fail to present an important bit of information. Remember, *you close the interview yourself.*

The chairman will then say, "That is all, Mr. ______, thank you." Do not be startled; the interview is over, and quicker than you think. Thank him, gather your belongings and take your leave. Save your sigh of relief for the other side of the door.

How to put your best foot forward

Throughout this entire process, you may feel that the board individually and collectively is trying to pierce your defenses, seek out your hidden weaknesses and embarrass and confuse you. Actually, this is not true. They are obliged to make an appraisal of your qualifications for the job you are seeking, and they want to see you in your best light. Remember, they must interview all candidates and a non-cooperative candidate may become a failure in spite of their best efforts to bring out his qualifications. Here are 15 suggestions that will help you:

1) Be natural – Keep your attitude confident, not cocky

If you are not confident that you can do the job, do not expect the board to be. Do not apologize for your weaknesses, try to bring out your strong points. The board is interested in a positive, not negative, presentation. Cockiness will antagonize any board member and make him wonder if you are covering up a weakness by a false show of strength.

2) Get comfortable, but don't lounge or sprawl

Sit erectly but not stiffly. A careless posture may lead the board to conclude that you are careless in other things, or at least that you are not impressed by the importance of the occasion. Either conclusion is natural, even if incorrect. Do not fuss with your clothing, a pencil or an ashtray. Your hands may occasionally be useful to emphasize a point; do not let them become a point of distraction.

3) Do not wisecrack or make small talk

This is a serious situation, and your attitude should show that you consider it as such. Further, the time of the board is limited – they do not want to waste it, and neither should you.

4) Do not exaggerate your experience or abilities

In the first place, from information in the application or other interviews and sources, the board may know more about you than you think. Secondly, you probably will not get away with it. An experienced board is rather adept at spotting such a situation, so do not take the chance.

5) If you know a board member, do not make a point of it, yet do not hide it

Certainly you are not fooling him, and probably not the other members of the board. Do not try to take advantage of your acquaintanceship – it will probably do you little good.

6) Do not dominate the interview

Let the board do that. They will give you the clues – do not assume that you have to do all the talking. Realize that the board has a number of questions to ask you, and do not try to take up all the interview time by showing off your extensive knowledge of the answer to the first one.

7) Be attentive

You only have 20 minutes or so, and you should keep your attention at its sharpest throughout. When a member is addressing a problem or question to you, give him your undivided attention. Address your reply principally to him, but do not exclude the other board members.

8) Do not interrupt

A board member may be stating a problem for you to analyze. He will ask you a question when the time comes. Let him state the problem, and wait for the question.

9) Make sure you understand the question

Do not try to answer until you are sure what the question is. If it is not clear, restate it in your own words or ask the board member to clarify it for you. However, do not haggle about minor elements.

10) Reply promptly but not hastily

A common entry on oral board rating sheets is "candidate responded readily," or "candidate hesitated in replies." Respond as promptly and quickly as you can, but do not jump to a hasty, ill-considered answer.

11) Do not be peremptory in your answers

A brief answer is proper – but do not fire your answer back. That is a losing game from your point of view. The board member can probably ask questions much faster than you can answer them.

12) Do not try to create the answer you think the board member wants

He is interested in what kind of mind you have and how it works – not in playing games. Furthermore, he can usually spot this practice and will actually grade you down on it.

13) Do not switch sides in your reply merely to agree with a board member

Frequently, a member will take a contrary position merely to draw you out and to see if you are willing and able to defend your point of view. Do not start a debate, yet do not surrender a good position. If a position is worth taking, it is worth defending.

14) Do not be afraid to admit an error in judgment if you are shown to be wrong

The board knows that you are forced to reply without any opportunity for careful consideration. Your answer may be demonstrably wrong. If so, admit it and get on with the interview.

15) Do not dwell at length on your present job

The opening question may relate to your present assignment. Answer the question but do not go into an extended discussion. You are being examined for a *new* job, not your present one. As a matter of fact, try to phrase ALL your answers in terms of the job for which you are being examined.

Basis of Rating

Probably you will forget most of these "do's" and "don'ts" when you walk into the oral interview room. Even remembering them all will not ensure you a passing grade. Perhaps you did not have the qualifications in the first place. But remembering them will help you to put your best foot forward, without treading on the toes of the board members.

Rumor and popular opinion to the contrary notwithstanding, an oral board wants you to make the best appearance possible. They know you are under pressure – but they also want to see how you respond to it as a guide to what your reaction would be under the pressures of the job you seek. They will be influenced by the degree of poise you display, the personal traits you show and the manner in which you respond.

ABOUT THIS BOOK

This book contains tests divided into Examination Sections. Go through each test, answering every question in the margin. We have also attached a sample answer sheet at the back of the book that can be removed and used. At the end of each test look at the answer key and check your answers. On the ones you got wrong, look at the right answer choice and learn. Do not fill in the answers first. Do not memorize the questions and answers, but understand the answer and principles involved. On your test, the questions will likely be different from the samples. Questions are changed and new ones added. If you understand these past questions you should have success with any changes that arise. Tests may consist of several types of questions. We have additional books on each subject should more study be advisable or necessary for you. Finally, the more you study, the better prepared you will be. This book is intended to be the last thing you study before you walk into the examination room. Prior study of relevant texts is also recommended. NLC publishes some of these in our Fundamental Series. Knowledge and good sense are important factors in passing your exam. Good luck also helps. So now study this Passbook, absorb the material contained within and take that knowledge into the examination. Then do your best to pass that exam.

EXAMINATION SECTION

EXAMINATION SECTION
TEST 1

DIRECTIONS: Each question or incomplete statement is followed by several suggested answers or completions. Select the one that BEST answers the question or completes the statement. *PRINT THE LETTER OF THE CORRECT ANSWER IN THE SPACE AT THE RIGHT.*

1. Before a new hog bristle paint brush is used for the first time with an oil paint, it should be soaked in 1.____

   A. water for two hours
   B. linseed oil overnight
   C. carbon tetrachloride overnight
   D. benzine for eight hours

2. A paint of good quality wears away slowly by 2.____

   A. chalking B. checking
   C. cracking D. scaling

3. The mixing of paints by pouring back and forth from one container to another is called 3.____

   A. blooming B. blushing
   C. boxing D. bubbling

4. If a roller is to be used the next day and you do not want to clean the roller, the paint can be prevented from drying on the roller by wrapping it 4.____

   A. tightly in tissue paper B. loosely in a clean rag
   C. loosely in newspaper D. tightly in aluminum foil

5. To prevent slipping, the bottom of ladder side rails are often fitted with 5.____

   A. locks B. shoes C. stirrups D. wedges

6. For removing foreign matter from surfaces prior to painting, 6.____

   A. solvent cleaning is satisfactory for use on concrete or masonry
   B. acid cleaning is not satisfactory for use on concrete or masonry
   C. alkali cleaning is satisfactory for use on aluminum or stainless steel
   D. acid cleaning is not satisfactory for use on aluminum or stainless steel

7. A structural steel paint job will last longer if the dry coats of paint are 7.____

   A. rigid and thin B. elastic and thick
   C. rigid and thick D. elastic and thin

8. A material that is added to a paint vehicle to regulate its consistency, increase its spreading power, and facilitate its application is called a(n) 8.____

   A. drier B. extender C. activator D. thinner

Questions 9-13.

DIRECTIONS: Questions 9 through 13, inclusive, refer to Column I and Column II below. For each type of paint or coating listed in Column I, select the proper solvent to use from the list in Column II.

COLUMN I

9. Calcimine 9.____
10. Latex 10.____
11. Portland cement 11.____
12. Shellac 12.____
13. Vinyl resin 13.____

COLUMN II

A. denatured alcohol
B. turpentine
C. red lead
D. water

Questions 14-18.

DIRECTIONS: Questions 14 through 18, inclusive, refer to Column I and Column II below. For each type of paint deterioration listed in Column I, select the statement in Column II which BEST describes the cause of this type of deterioration.

COLUMN I

14. Alligatoring 14.____
15. Blistering 15.____
16. Checking 16.____
17. Crazing 17.____
18. Wrinkling 18.____

COLUMN II

A. the coat applied was too thick
B. the bonding between the finish coating and the primer is poor
C. the covering of a relatively soft coat by a relatively hard one
D. water seeping from base surface pushing off the paint film

19. One of the precautions that should be taken in using and storing manila rope in order to prolong its life is to 19.____

A. avoid whipping the ends of the rope
B. avoid knotting the rope unnecessarily
C. make sure to store the rope in a warm, damp location
D. coil the rope counter-clockwise on the floor

20. When adding tinting colors, colors-in-oil CANNOT be used with ______ paints. 20.____

A. water-thinned
B. oil-based
C. alkyd resin
D. chlorinated rubber

21. Green lumber should NOT be used for scaffolding because it 21.____

A. is too expensive
B. may warp upon drying
C. will not hold nails well
D. splits easily when nailed

22. Combustible materials should be stored in a(n) ______ container. 22.____

A. open metal
B. open plastic
C. covered metal
D. covered plastic

23. When storing paint materials, the temperature range of the storage area ideally should be between 23.____

A. 20° F and 40° F
B. 40° F and 65° F
C. 65° F and 85° F
D. 85° F and 100° F

24. Of the sizes given, the COARSEST grade of sandpaper is number 24.____

A. 1
B. 2
C. 00
D. 0

25. Of the following, the one which is flammable is 25.____

A. acetone
B. carbon tetrachloride
C. methylene chloride
D. sodium bicarbonate

KEY (CORRECT ANSWERS)

1.	B	11.	D
2.	A	12.	A
3.	C	13.	B
4.	D	14.	C
5.	B	15.	D
6.	D	16.	B
7.	D	17.	B
8.	D	18.	A
9.	D	19.	B
10.	D	20.	A

21. B
22. C
23. C
24. B
25. A

TEST 2

DIRECTIONS: Each question or incomplete statement is followed by several suggested answers or completions. Select the one that BEST answers the question or completes the statement. *PRINT THE LETTER OF THE CORRECT ANSWER IN THE SPACE AT THE RIGHT.*

1. Applying paint which has NOT been thoroughly mixed or stirred may result in 1.____

   A. alligatoring B. cracking
   C. running D. scaling

2. In painting terminology, a holiday is a 2.____

   A. jelly-like substance B. non-reactive pigment
   C. skip D. purifier

3. The erosion of a paint film after rapid chalking is called 3.____

   A. livering B. separating
   C. skinning D. washing

4. When painting an interior plywood wall surface with a roller, one should NOT 4.____

   A. reverse direction
   B. spin the roller at the end of each stroke
   C. feather out final strokes to pick up any excess paint on the surface
   D. roll paint onto the surface, working from the dry area into the just painted area

5. The deposit of soluble white salts on the surface of brick and other masonry is called 5.____

   A. efflorescence B. elasticity
   C. erosion D. extending

6. Rope that has been used near acid, caustic or chemical fumes should be 6.____

   A. disposed of immediately
   B. given a visible check before re-using
   C. tested for strength before using again
   D. stored and not used again for at least one year

7. In order to make a room appear larger, it should be painted with a 7.____

   A. design on at least one wall
   B. dark colored paint on all walls
   C. light colored paint on all walls
   D. dark color on two walls and a light color on the other walls

8. Before a cement-water paint is applied to brick or concrete surfaces, these surfaces should be 8.____

   A. thoroughly dampened B. completely dry
   C. varnished D. shellacked

9. A prime objective for a foreman is to give the best paint job for the least money. This should be evaluated by calculating the length of time between repaintings and also the original expense. 9._____
With this in mind, the MOST economical paint job is one which will have an original expense of

A. $15,000 and last four years
B. $10,000 and last three years
C. $8,000 and last two years
D. $5,000 and last one year

10. While waiting for an ambulance to take away a painter who has fallen off a ladder and broken his leg, a foreman should 10._____

A. get a board and use it as a temporary splint
B. apply a tourniquet with whatever is available
C. make the painter as comfortable as possible without moving him
D. make an emergency stretcher and have another employee help place the painter on it

11. A certain agency has decided to do its own painting and employ a permanent staff of painters. 11._____
This staff should be large enough to

A. train new employees
B. respond to seasonal demands
C. handle the regular day-to-day work load
D. handle peak work loads resulting from emergency jobs

12. The practice of employing an outside contractor's painters on the same job with painters of a public agency is GENERALLY a 12._____

A. *poor* practice because it is not safe
B. *good* practice because it is more efficient
C. *poor* practice because of potential labor problems
D. *good* practice because it will insure consistent quality

13. When using an outside contractor, it is of prime importance to define the scope of his work. 13._____
The MOST logical reason for this practice is that it will

A. reduce job-related accidents
B. help prevent costly misunderstandings
C. reduce the contractor's labor problems
D. help make a good impression on the contractor

14. In order to avoid excessive cost to an agency on a cost-plus painting contract, it is MOST important for the agency to give the contractor 14._____

A. adequate manpower
B. adequate paint and supplies
C. continuous direction in scheduling the work
D. continuous direction in departmental accounting procedures

15. The sum of the following dimensions: 3'2 1/4", 0'8 7/8", 2'6 3/8", 2'9 3/4", and 1'0" is 15.____

A. 9'2 7/8"
B. 10'3 1/4"
C. 10'7 3/8"
D. 11'4 1/4"

16. If the scale of a drawing is 1/8" to the foot, then a 1/2" measurement on the drawing would represent an actual length of ______ feet. 16.____

A. 2
B. 4
C. 8
D. 16

Questions 17-21.

DIRECTIONS: Questions 17 through 21, inclusive, are based on the information given below.

A crew of 5 painters are going to paint 55 rooms. They will be painting only the walls, which are all 10 feet high. The rooms have the following dimensions: 30 rooms are 25 feet long and 15 feet wide and the remaining rooms are 20 feet long and 15 feet wide. All walls will be painted the same color and will require 2 coats. Coverage is 500 square feet per gallon. Each painter can cover 700 square feet of wall per day and works 7 hours per day.

17. Assume that 20% of the total wall surface consists of windows which are not to be painted. The total wall surface, in square feet, to be painted is *most nearly* 17.____

A. 15,600
B. 21,800
C. 33,200
D. 41,500

18. Assume the total wall surface to be painted is 49,500 square feet per coat of paint. The total number of gallons of paint needed for a complete job is *most nearly* 18.____

A. 250
B. 200
C. 150
D. 100

19. The total number of working days required for this crew to cover 49,500 square feet of wall surface with two coats of paint is *most nearly* 19.____

A. 17
B. 23
C. 29
D. 35

20. Assuming each painter earns $11.20 per hour, the total cost in painter's wages for a job which takes 21 working days to complete is *most nearly* 20.____

A. $10,460
B. $8,240
C. $6,020
D. $4,720

21. If two painters are sick for two days each and they are not replaced, the total time to complete this job would be extended approximately ______ day(s). 21.____

A. 1
B. 2
C. 3
D. 4

22. *Boxing* of paint refers to the ______ of paints. 22.____

A. mixing
B. storage
C. use
D. canning

23. The vehicle used with latex paints is USUALLY 23.____

A. linseed oil
B. shellac
C. varnish
D. water

24. When painting wood, nail holes should be puttied 24.____

A. before applying the prime coat
B. after applying the prime coat but before the second coat
C. after applying the second coat but before the third coat
D. after applying the third coat

25. The pigment MOST often used in a prime coat of paint on steel to prevent rusting is 25.____

A. lampblack
B. calcimine
C. zinc oxide
D. red lead

KEY (CORRECT ANSWERS)

1.	C	11.	C
2.	C	12.	C
3.	D	13.	B
4.	B	14.	C
5.	A	15.	B
6.	A	16.	B
7.	C	17.	C
8.	A	18.	B
9.	B	19.	C
10.	C	20.	B

21. A
22. A
23. D
24. B
25. D

EXAMINATION SECTION
TEST 1

DIRECTIONS: Each question or incomplete statement is followed by several suggested answers or completions. Select the one that BEST answers the question or completes the statement. *PRINT THE LETTER OF THE CORRECT ANSWER IN THE SPACE AT THE RIGHT.*

1. Creosote stain 1.____

   A. is a poor preservative
   B. is an excellent primer for any paint
   C. will usually bleed through most paints
   D. should be well-weathered before paint is applied over it

2. *Checking* of a paint coat is *most likely* to occur if 2.____

   A. the top coat is too soft
   B. the undercoat is too soft
   C. there is too much drying time between coats
   D. water is present under the paint

3. Of the following woods, the one that will retain paint the LONGEST assuming similar conditions of weathering is 3.____

   A. Douglas fir
   B. hemlock
   C. redwood
   D. spruce

4. Of the following types of fabric rollers, the one that should NOT be used to apply latex paint is a _____ roller. 4.____

   A. lambs wool
   B. dynel
   C. mohair
   D. dacron

5. A zinc chromate priming paint would *most likely* be used on 5.____

   A. concrete B. metal C. plaster D. wood

6. Normal paint deterioration on the exterior wood of a building is usually MOST severe on the side facing 6.____

   A. east B. north C. south D. west

7. A 2" brush should have bristles that are *most nearly* _____ long. 7.____

   A. 1" B. 2" C. 3" D. 4"

8. Discolored streaks are often formed on painted surfaces which are below unpainted metal parts of a building. 8.____
   Of the following metals, the one that will cause the LEAST discoloration of a painted surface is

   A. copper B. iron C. steel D. aluminum

9. Of the following, the difference between glazing compound and putty is that glazing compound 9.____

   A. is applied directly to unprimed wood
   B. is used only on metal sash
   C. should be painted before it is dry
   D. does not dry hard throughout

10. Of the following materials, the one that is NOT considered a thinner for some type of paint is 10.____

    A. linseed oil
    B. resin
    C. turpentine
    D. water

11. Skinning would probably be MOST troublesome when the paint is a 11.____

    A. gloss enamel
    B. varnish paint
    C. latex paint
    D. stain

12. Of the following pigments, the one that is LEAST likely to be used in a ready-mixed paint is 12.____

    A. aluminum
    B. iron oxide
    C. lithopone
    D. red lead

13. Extenders added to paints are of LEAST value in 13.____

    A. adjusting the consistency of the paint
    B. adding to the hiding power of the paint
    C. controlling the gloss of the paint
    D. increasing the durability of the paint

14. The type of paint that is BEST for application to a new concrete interior wall is 14.____

    A. enamel
    B. latex
    C. oil-base
    D. varnish

15. In comparing spray painting to applying paint with a brush on woodwork, it is CORRECT to say that spray painting 15.____

    A. requires a cleaner surface
    B. is always less time-consuming
    C. generally gives better results
    D. requires more skill

16. Of the following woods, the one that is LEAST likely to be used for exterior building surfaces is 16.____

    A. cedar
    B. oak
    C. pine
    D. spruce

17. Stages in the normal deterioration of paint on exterior woodwork include chalking, checking, and flatting. 17.____
    The ORDER in which these stages usually take place is

    A. chalking, checking, flatting
    B. chalking, flatting, checking
    C. flatting, checking, chalking
    D. flatting, chalking, checking

18. Wrinkling of paint on exterior woodwork is LIKELY to be caused by 18.____

A. damp weather
B. the presence of oily dirt on the surface before painting
C. a sharp drop in temperature after the paint is applied
D. too much thinning of the paint

19. The height of the nap on a roller used in painting chain-link fences should be *most nearly* 19.____

A. 1/4" B. 3/8" C. 3/4" D. 1 1/4"

20. In three-coat work on exterior woodwork, the FEWEST number of gallons of paint will be required for the 20.____

A. priming coat
B. second coat (first coat of finish paint)
C. third coat
D. coat with the deepest color

21. Under favorable conditions, the drying time between coats of a linseed oil exterior paint should be AT LEAST ____ hours. 21.____

A. 12 B. 24 C. 48 D. 96

22. An oak floor is to be filled, stained, and varnished. The proper SEQUENCE is 22.____

A. fill, stain, varnish
B. varnish, stain, fill, varnish
C. stain, fill, varnish
D. stain, varnish, fill, varnish

23. When varnishing a floor, POOR results will be obtained if 23.____

A. several smooth, thin coats of varnish are used
B. varnish is first brushed across the grain and then leveled off with the grain
C. very fine sandpaper or steel wool is used between coats
D. any coat is applied on top of a coat that is not thoroughly dry

24. When painting of exterior woodwork must be done in near-freezing weather, 24.____

A. the paint should be thinned with linseed oil
B. driers should be added to the paint
C. the paint should be warmed before use
D. the drying time between coats should be reduced

25. The total thickness of a two-coat system used for exterior woodwork, as compared to the total coating thickness of three-coat work, should be 25.____

A. about the same
B. greater
C. less
D. either greater or less depending on the type of pigment

26. When new softwood steps are to be painted, the BEST protection and appearance will be provided by 26.____

A. many thick coats B. many thin coats
C. two or three thick coats D. two or three thin coats

27. A second coat of paint should be applied ONLY when the first 27.____

A. no longer appears to be wet
B. is tacky to the touch
C. is dry to the touch
D. can be sandpapered without gumming

28. The BEST way to thin varnish with turpentine is to add the turpentine ____ while stirring the varnish ____. 28.____

A. quickly; vigorously
B. quickly; gently
C. gradually; vigorously
D. gradually; gently

29. The bristles of a 3-inch brush are 4 inches long. The depth to which the bristles should be dipped when picking up paint is *usually* ____ inch(es). 29.____

A. 1
B. 2
C. 3
D. 4

30. The BEST way to apply the first coat of an oil paint to exterior wood siding is to 30.____

A. empty the brush in one long horizontal stroke, then work the paint in with short vertical strokes
B. use short vertical, overlapping strokes until brush is empty, then work in with long horizontal strokes
C. use short, light strokes, flowing the paint on with the least possible brushing
D. apply the paint in spots, brush out each spot into its neighbor, and then smooth the entire strip with long, gentle strokes

31. A pan-type roller is to be used to apply latex paint to plaster. 31.____
The statement that gives the MOST complete description of the proper way to load the roller is:

A. Roll it no deeper than the thickness of the cover in the shallow section of a half-filled tray
B. Roll it no deeper than the thickness of the cover in the shallow section of a half-filled tray, then roll it over the corrugations on the exposed bottom of the tray
C. Roll it from one end of a well-filled tray to the other
D. Place it in the deepest portion of the tray so that the cover can soak up an ample paint supply

32. A paneled door is to be painted white. 32.____
The parts of the door which should be painted FIRST are the

A. moulded edges of the panels
B. panels
C. horizontal crossboards
D. vertical sideboards

33. Of the following, the component of an oleo-resinous varnish that is NOT found in a spirit varnish is 33.____

A. resin
B. volatile solvent
C. drying oil
D. shellac

34. Mercuric chloride (also known as bichloride of mercury and corrosive sublimate) is added to paint to PREVENT 34.____

A. bleeding
B. fading
C. flatting
D. mildewing

35. Boiled linseed oil differs from raw linseed oil PRINCIPALLY in that it 35.____

A. has been heated to high temperatures
B. has had solvent added to it
C. has had air blown through it
D. is acid refined

36. A white lead paint becomes too thick in cold weather. 36.____
The BEST way of restoring the paint to its proper consistency is to add

A. alcohol
B. drier
C. linseed oil
D. turpentine

37. Refined lac dissolved in denatured alcohol is called 37.____

A. lacquer B. sealer C. shellac D. varnish

38. Adding titanium dioxide to a lead-zinc paint would probably yield a paint with all of the following characteristics EXCEPT 38.____

A. greater durability
B. hardened film
C. increased hiding power
D. reduced chalking

39. The BEST varnish for a wood floor is a ____ varnish. 39.____

A. short-oil
B. medium-oil
C. fish oil
D. latex

40. Of the following paints, the one which contains the GREATEST percentage of pigment is 40.____

A. flat oil paint
B. semi-gloss oil paint
C. gloss oil paint
D. enamel

KEY (CORRECT ANSWERS)

1.	C	11.	A	21.	C	31.	B
2.	B	12.	C	22.	C	32.	A
3.	C	13.	B	23.	D	33.	C
4.	A	14.	B	24.	C	34.	D
5.	B	15.	A	25.	A	35.	A
6.	C	16.	B	26.	D	36.	D
7.	C	17.	D	27.	D	37.	C
8.	D	18.	C	28.	D	38.	B
9.	D	19.	D	29.	B	39.	B
10.	B	20.	B	30.	D	40.	A

TEST 2

DIRECTIONS: Each question or incomplete statement is followed by several suggested answers or completions. Select the one that BEST answers the question or completes the statement. *PRINT THE LETTER OF THE CORRECT ANSWER IN THE SPACE AT THE RIGHT.*

1. The aluminum of a window frame has oxidized. Before painting the frames, the oxide 1.____

   A. should be removed with acid
   B. should be sealed with varnish
   C. should be removed by wire brushing
   D. may be left undisturbed

2. A dirty wall surface that is to be repainted should be washed starting at 2.____

   A. the bottom
   B. the top
   C. either the top or the bottom, whichever is easier
   D. the bottom with plaster walls and at the top with brick walls

3. The PROPER procedure to follow in connection with the priming of wood siding is that 3.____

   A. knots should be shellacked after priming
   B. knots should be shellacked before priming
   C. nail holes should be puttied before priming
   D. wood siding is never back primed

4. The one of the following statements concerning the repainting of interior wood that is CORRECT is: 4.____

   A. Worn spots should be spotprimed and sandpapered lightly after the primer has dried
   B. Greasy dirt should be wiped off with a dry cloth
   C. There is no objection to applying the paint over old varnish
   D. Yearly painting is better than less frequent painting

5. When an exterior paint that is chalking is rubbed lightly with a finger, the paint comes off right down to the bare wood. 5.____
   The one of the following statements that is CORRECT concerning this condition is:

   A. This is normal and no repainting is necessary
   B. The surface should be dusted and then repainted
   C. The surface should be scraped or wire-brushed and then repainted
   D. This indicates mildew. The surface should be washed with water and then repainted

6. A new concrete wall that was cast against smooth metal forms is to be painted with a water-cement paint. Preparation of the wall should include all of the following EXCEPT 6.____

   A. neutralization of alkalies
   B. removal of oil
   C. roughening of surface
   D. removal of efflorescence

7. A new concrete wall is to be painted with an oil-base paint. 7.____
Before painting, it is BEST to treat the surface of the concrete with a solution of

A. lime
B. beatsal
C. washing soda and trisodium phosphate
D. phosphoric acid and zinc chloride

8. Two coats of masonry filler are applied to a cinder block wall before it is painted. 8.____
The second coat should be allowed to dry for a MINIMUM of ____ hours.

A. 6 B. 12 C. 18 D. 24

9. A strip about 1 inch wide and over a foot long has been gouged out of a plaster and metal lath wall. The depth of gouging varies but does not exceed 1/4 inch. 9.____
Before filling this area with patching plaster, the area should be

A. cut out down to the metal lath
B. cut out to a uniform depth of 1/4 inch
C. washed with turpentine
D. painted with a sealer

10. When cracks in concrete are repaired by using portland cement grout, hydrated lime is often added to the grout in order to 10.____

A. increase the strength of the grout
B. slow up the curing of the grout
C. change the color of the grout
D. make the grout expand

11. Exterior woodwork that is to be repainted has some spots that are still glossy. 11.____
The spots should

A. not be repainted
B. be coated very thinly
C. be dulled by sanding or washing with a paint cleaning compound
D. be treated exactly like the rest of the surface

12. There are traces of wax on interior woodwork that is to be repainted. 12.____
After wiping the surface with turpentine, it is BEST practice to

A. apply the new paint immediately
B. wash the woodwork with a trisodium phosphate solution
C. allow the wood to dry before dusting and repainting
D. seal the woodwork and then stain it

13. Of the following colors, the one that is NOT a primary color is 13.____

A. blue B. green C. red D. yellow

14. The total number of primary and secondary colors is 14.____

A. 4 B. 6 C. 9 D. 12

15. A small amount of orange paint is added to a bright blue paint. The color of the mixture is 15.____

A. blue-green
B. purple
C. violet
D. a less bright blue

16. A mixture of blue and yellow paints produces a _____ paint. 16.____

A. cream
B. green
C. orange
D. violet

17. The complementary color of any warm color is a _____ color. 17.____

A. secondary
B. tertiary
C. cool
D. warm

18. The pigments that could be mixed together to make a dark green paint are lampblack, prussian blue, and 18.____

A. burnt sienna
B. chromium hydroxide
C. lemon chrome yellow
D. Venetian red

19. The color of burnt umber is 19.____

A. orange
B. purple
C. blue
D. red-yellow

20. A color is viewed in sunlight and then in incandescent light. The color will appear to 20.____

A. acquire a greenish tinge in the incandescent light
B. acquire a reddish tinge in the incandescent light
C. acquire a whitish tinge in the fluorescent light
D. be unchanged in the artificial light

21. Of the following requirements for the bristles of a paint brush, the one that is LEAST important is that the bristles 21.____

A. be flagged
B. be tapered
C. vary in length
D. be made of hog hair

22. A non-bleeder type gun for spray painting must be used with a(n) 22.____

A. air tank
B. compressor with two or more cylinders
C. external-mix nozzle
D. pressure feed

23. After selecting an air cap and a fluid tip for a spray gun, one of the BEST ways to balance air and paint flow is to 23.____

A. use the fluid-adjusting screw
B. thin or thicken the paint
C. grease the pattern-adjustment screw
D. jiggle the trigger as the gun is moved

24. A 45° nozzle should be used on a spray gun when painting 24.____

A. a wall at an inside corner
B. a wall at an outside corner
C. the ceiling of a room
D. the sides of cabinets

25. Parallel strokes, next to each other, from a spray gun should overlap 25.____

A. not at all
B. just enough to insure that there are no bare areas
C. ~~about 1/3 of the width of the stroke~~
D. 2/3 to 3/4 of the width of the stroke

26. Paint has dried in the nozzle of a spray gun. The BEST way to clean the nozzle is to 26.____

A. put clean thinner in the container and operate the gun
B. loosen the paint with a wire and blow it out with air
C. wipe the nozzle with a rag soaked in thinner
D. disassemble the nozzle and soak the parts in thinner

27. The BEST solvent to use to clean a brush that has been used to apply shellac is 27.____

A. denatured alcohol
B. lacquer thinner
C. raw linseed oil
D. turpentine

28. Bleached shellac would MOST probably be used 28.____

A. where a flexible coating is required
B. where resistance to liquids is required
C. on a light-colored wood
D. on exterior woodwork

29. Of the following conditions, the one that is LEAST likely to indicate that a fiber rope is unsafe is 29.____

A. broken fibers on the outside
B. strands have begun to unlay
C. dirt on the inside
D. dirt on the outside

30. The CHIEF purpose of the stirrups sometimes attached to a boatswain's chair is to 30.____

A. allow the occupant to reach a greater area from any one position of the chair
B. keep the occupant from falling out
C. keep the occupant's feet from *going to sleep*
D. allow the occupant to control the swinging of the chair

31. The BEST knot to use to make a comfortable sling to set in is a 31.____

A. running bowline
B. double bowline
C. slip knot
D. cat's paw or rocking hitch

32. The BEST life line is made of 32.____

A. manila rope
B. nylon rope
C. wire rope with a hemp center
D. wire rope with a wire rope center

33. Of the following conditions, the one that is the MOST important requirement in the storage of manila rope is 33.____

A. good air circulation
B. exposure to sunlight
C. exposure to a source of heat
D. ~~high humidity~~

34. A thimble would *most likely* be used when rope is 34.____

A. made into a grommet
B. unknotted
C. reeved
D. attached to a ring

35. The factor of safety (ratio of breaking strength to safe load) recommended for manila rope is *most nearly* 35.____

A. 3 B. 5 C. 9 D. 11

36. A tackle has two triple pulley blocks. It is used to raise a load to the roof of a building with the lead line being pulled by men on the ground. 36.____
To raise the load one foot, the lead line must move ____ feet.

A. 3 B. 4 C. 6 D. 8

37. Vapor from a solvent in a closed space is 37.____

A. explosive regardless of the concentration
B. explosive for all concentrations exceeding a certain percentage
C. not explosive if the concentration exceeds a certain percentage
D. not explosive under any conditions

38. With respect to the erection and use of swinging scaffolds raised and lowered with block and tackle, it is CORRECT to say that 38.____

A. life lines should be so arranged that if a man falls, the line will hang from a point no lower than the level he was working on
B. the hook of the lower pulley block should be moused
C. when two scaffolds are used end-to-end on a wall, they should be securely lashed together
D. window cleaners' anchors may be used for tie-ins

39. With respect to the erection and use of movable scaffold mounted on wheels, it is CORRECT to say that 39.____

A. men on the scaffold must stand still while it is being moved
B. the working platform must have a guard rail on three sides
C. casters or wheels should be securely chocked with hardwood blocks when the scaffold is in use
D. a ladder used to reach the working platform must be fastened to the scaffold

40. For proper care, wood ladders should be 40.____

A. painted at regular intervals
B. coated with varnish, shellac, or linseed oil
C. stored upright by leaning against a wall
D. stored horizontally with supports at the ends only

KEY (CORRECT ANSWERS)

1.	D	11.	C	21.	D	31.	B
2.	A	12.	B	22.	A	32.	B
3.	A	13.	B	23.	A	33.	A
4.	A	14.	B	24.	C	34.	D
5.	C	15.	D	25.	C	35.	B
6.	A	16.	B	26.	D	36.	C
7.	D	17.	C	27.	A	37.	C
8.	C	18.	C	28.	C	38.	A
9.	A	19.	D	29.	D	39.	D
10.	B	20.	B	30.	C	40.	B

EXAMINATION SECTION
TEST 1

DIRECTIONS: Each question or incomplete statement is followed by several suggested answers or completions. Select the one that BEST answers the question or completes the statement. *PRINT THE LETTER OF THE CORRECT ANSWER IN THE SPACE AT THE RIGHT.*

1. Lacquer thinner would *most likely* be used to 1.____

   A. clean oil paint from a brush immediately after use
   B. clean a paint brush upon which paint has hardened
   C. rinse a new paint brush before using it
   D. remove paint from the hands

2. Before repainting a glossy surface, the surface should be sanded to prevent 2.____

   A. crawling B. blistering
   C. alligatoring D. mildewing

3. Raw linseed oil with a drier added could *best* be substituted for 3.____

   A. turpentine B. varnish
   C. mineral spirits D. boiled linseed oil

4. A white lead paste consists *mainly* of white lead *and* 4.____

   A. turpentine B. linseed oil
   C. water D. varnish

Questions 5 - 7.

Questions 5 through 7, inclusive, refer to the paint formulas which follow. These paints are to be used for a three-coat job on exterior new wood.

	Formula #1	Formula #2	Formula #3
White lead - lb.	100	100	100
Raw linseed oil - gal.	1 1/2	3 1/4	4
Turpentine - gal.	1 1/2	0	2
Liquid drier - pt.	1	1	1

5. The priming coat should be Formula # ______. 5.____

   A. 1 B. 1 or 2 C. 2 D. 3

6. The body coat should be Formula # ______. 6.____

   A. 1 B. 2 C. 1 or 2 D. 3

7. The finish coat should be Formula # ______. 7.____

   A. 1 B. 2 C. 2 or 3 D. 3

8. The fire escapes of an apartment building are to be given a three-coat paint job. The difference between the priming coat and the body coat would, most probably be, in the 8.____

   A. vehicle B. pigment
   C. thinner D. drier

9. A polishing wax consists of carnauba wax in an organic solvent. This wax polish should NOT be used on 9.____

A. furniture
B. hardwood floors
C. softwood floors
D. rubber-tile floors

10. The "length" of a varnish refers to the 10.____

A. area covered per gallon
B. period of time it will last without deterioration
C. ratio of oil to resin
D. type of resin used

11. Of the following coatings, the *one* which is *most likely* applied by spraying rather than brushing is 11.____

A. oil paint
B. lacquer
C. shellac
D. varnish

12. A solvent in a paint is classified as a 12.____

A. pigment
B. vehicle
C. drying oil
D. binder

13. A pigmented coating or paint is preferred to an unpigmented coating when the coating is exposed to 13.____

A. wind
B. sun
C. rain
D. wide temperature changes

14. Putty used for glazing wood sash should be applied ______ coat. 14.____

A. *before* the prime
B. *after* the finish
C. *after* the prime
D. *after* the body

15. Zinc oxide added to an exterior white oil paint reduces 15.____

A. chalking
B. checking and cracking
C. hiding power
D. film hardness

16. The length of the bristles of a paint brush, which has a width of 4 inches, should be *at least* ______ inches. 16.____

A. 2
B. 3
C. 3 1/2
D. 4

17. In painting a wooden wall with a brush, the paint should be laid on with brush strokes in one direction and then smoothed with brush strokes at an angle to the lay-on strokes. The smoothing strokes should be 17.____

A. parallel to the grain
B. perpendicular to the grain
C. vertical, regardless of grain
D. horizontal, regardless of grain

18. Painting wood with a brush is better than using a spray gun *primarily* because brushing 18.____

A. is cheaper
B. is faster
C. ~~works the paint into the pores of the wood~~
D. uses less paint

19. *Before* putty that has been used in glazing wood sash can be painted, it should be allowed to dry for *at least* 19.____

A. 4 hours
B. 24 hours
C. 48 hours
D. 4 days

20. Interior wood trim in a plasterboard room in which the plaster is not to be painted *should be* painted 20.____

A. *as soon as* the plasterers have finished
B. *as soon as* the plaster is hard
C. *when* the plaster is relatively dry
D. *only after* the plaster is equally moist and dry

21. Shellac, used as a sealer on knots and pitch streaks, should be allowed to dry for *at least* 21.____

A. 4 hours
B. 24 hours
C. 48 hours
D. one week

22. Of the following ready-mixed coatings, the *one* which is *least likely* to require re-mixing before use is 22.____

A. exterior oil paint
B. flat finish paint
C. semi-gloss enamel
D. varnish

23. Failure of exterior paint is *usually* most pronounced on the following side or sides of a building: 23.____

A. North
B. South
C. East and West
D. North and South

24. A painter would *most likely* use builder's acid (muriatic acid and water) before painting 24.____

A. wood
B. steel
C. plastic
D. brickwork

25. A painter pours a little turpentine into a half-full can of enamel in such a way that the turpentine forms a thin layer on top of the enamel. 25.____
He does this, *most probably,* to

A. thin the enamel
B. prevent skinning
C. make the enamel more glossy
D. make the enamel less glossy

26. A brush which is used daily for the application of varnish has been suspended overnight in raw linseed oil. Before using, it is good practice to press the excess oil out of the brush *and* 26.____

A. use the brush without further treatment
B. rinse in clear water
C. rinse in alcohol
D. rinse in turpentine

27. Assume that a pigmented paint is stored in one-gallon cans and that it will not be used for at least one year. To insure that the paint will be usable when needed, the *usual* practice is to, occasionally, 27.____

A. open the cans and mix throughly
B. place the cans in a shaker
C. turn the cans over end-for-end
D. roll the cans and replace in the same position

28. Slimy matter that separates from some oils when they stand for a long time is *usually* known as 28.____

A. foots
B. efflorescence
C. gum
D. resin

29. "Spreading rate" is *usually* expressed in 29.____

A. square feet per hour
B. gallons per hour
C. square feet per gallon
D. gallons per square foot

30. The MOST viscous of the following fluids is 30.____

A. boiled linseed oil
B. raw linseed oil
C. turpentine
D. alcohol

31. Of the following ingredients, the *one* which would be classified as a drying oil is 31.____

A. turpentine
B. mineral spirits
C. crude oil
D. linseed oil

32. A concrete wall is to be painted using a cement-water paint. The paint is BEST applied by a 32.____

A. spray gun
B. white-wash brush
C. stiff-fiber brush
D. soft-fiber brush

33. The *principal* ingredient in white-wash is 33.____

A. chalk
B. lime paste
C. whiting
D. gypsum

34. Bichloride of mercury solution is NOT widely used to wash mildewed surfaces because 34.____

A. trisodium phosphate solution is more efficient
B. ~~it is expensive~~
C. it is a deadly poison
D. it deteriorates rapidly

35. A can of paint is to be mixed before using. 35.____
The use of a shaker to do the mixing is NOT recommended if the paint is

A. enamel
B. white lead outside wood paint
C. red lead paint for steel
D. deck paint

KEY (CORRECT ANSWERS)

1.	B	16.	D
2.	A	17.	A
3.	D	18.	C
4.	B	19.	C
5.	D	20.	C
6.	A	21.	A
7.	B	22.	D
8.	B	23.	B
9.	D	24.	D
10.	C	25.	B
11.	B	26.	D
12.	B	27.	C
13.	B	28.	A
14.	C	29.	C
15.	A	30.	A

31. D
32. C
33. B
34. C
35. A

TEST 2

DIRECTIONS: Each question or incomplete statement is followed by several suggested answers or completions. Select the one that BEST answers the question or completes the statement. *PRINT THE LETTER OF THE CORRECT ANSWER IN THE SPACE AT THE RIGHT.*

1. Weather conditions being the same, a paint is *most durable* when its color is 1._____

   A. white
   B. red
   C. green
   D. black

2. Of the following woods, the *one* which is POOREST in paint-holding quality is 2._____

   A. cedar
   B. southern yellow pine
   C. northern white pine
   D. spruce

3. A self-cleaning paint is one that 3._____

   A. checks
   B. alligators
   C. chalks
   D. scales

4. Blistering of paint on wood surfaces is *usually* caused by 4._____

   A. sunlight
   B. cold
   C. heat
   D. moisture

5. It is BEST to store paint in 5._____

   A. airtight rooms with windows
   B. airtight rooms without windows
   C. well-ventilated rooms with windows
   D. well-ventilated rooms without windows

6. Colors used in paints for painting concrete walls should be 6._____

   A. pigments in oil
   B. mineral pigments
   C. water colors
   D. tempera

7. Of the following, the BEST way to warm paint is to heat it 7._____

   A. in a hot water bath with the can lid loosened
   B. over an open flame with the can lid loosened
   C. in a hot water bath without loosening the can lid
   D. over an open flame without loosening the can lid

8. A rectangular wooden building occupies a ground space 27' 6" long by 18' 0" wide. The walls are 17' 6" high. Ignoring window and door spaces, the outside area requiring painting is, in square feet, most nearly, 8._____

   A. 1570
   B. 1590
   C. 1610
   D. 1630

9. The type of paint which should *only* be applied to a damp surface is 9._____

   A. oil paint
   B. lacquer
   C. cement paint
   D. aluminum paint

10. If blue and yellow paint are mixed together in equal proportions, the color of the mixture will be 10.____

A. red
B. violet
C. purple
D. green

11. A gray-colored paint may be made by mixing together, in proper proportions, the following colored paints: 11.____

A. red and yellow
B. yellow and white
C. white and blue
D. black and white

12. An orange-colored paint may be made by mixing together, in proper proportions, the following colored paints: 12.____

A. white and yellow
B. yellow and red
C. red and white
D. white and ivory

Questions 13-15.

Questions 13 to 15, inclusive, are based on the following statement:
Tints are made by adding pigment-in-oil to white lead paste.

13. If the pigment used is burnt sienna, the color of the tint will, most likely, be 13.____

A. brown
B. green
C. yellow
D. orange

14. If the pigment used is Venetian red, the color of the tint will, most likely, be 14.____

A. red
B. yellow
C. orange
D. blue

15. If the pigment used is chromium oxide, the color of the tint will, most likely, be 15.____

A. greenish blue
B. pale orange
C. yellowish green
D. yellowish orange

Questions 16-17.

Questions 16 and 17 are based on the following statement:
An exterior oil paint which is ready-mixed is to be used for a three-coat job.

16. The addition of one pint of raw linseed oil to one gallon of the paint would *normally* be made for the 16.____

A. priming coat
B. body coat
C. finish coat
D. priming and body coats

17. The addition of up to one pint of turpentine to one gallon of the paint would *normally* be made for the 17.____

A. priming coat
B. body coat
C. finish coat
D. priming and body coats

18. Assume that a certain type of cabinet can be painted in twenty minutes by using a brush or 15 min. with spraygun. Assuming that three painters are put on the job but only one spray gun is available, the number of hours required by the three painters to paint 100 cabinets, one using the spray gun and the other two using brushes, is, most nearly, 18.____

A. 9 B. 10 C. 11 D. 12

19. If it takes 5 painters 12 days to paint a building, the number of days it will take 9 painters to paint the same building, assuming all work is done at the same rate of speed, is, most nearly, 19.____

A. 5 1/2 B. 6 1/2 C. 7 1/2 D. 8 1/2

20. Assume that one gallon of paint, costing $6.25, is able to cover 500 square feet. If a painter can spread 1.25 gallons per day and receives $60.00 per day, the cost per 100 square feet for labor and paint for a one-coat application is, most nearly, 20.____

A. $12.00 B. $12.10 C. $12.20 D. $12.30

21. The strength of a swinging scaffold should be *at least* ______ the load it carries. 21.____

A. twice
B. 3 times
C. 4 times
D. 5 times

22. In order to maintain safety, *good* practice is that the maximum number of men that should be allowed on the ordinary swinging scaffold is 22.____

A. 4 B. 3 C. 2 D. 1

23. Muriatic acid is being used by a painter. The acid is *most likely* to do serious damage to the following part of a swinging scaffold: 23.____

A. metal stirrup or hangar
B. wooden floor boards
C. manila ropes
D. all of the foregoing

24. The *proper* size of a toe board on a swinging scaffold is, approximately, 24.____

A. 1" x 1"
B. 2" x 2"
C. 2" x 4"
D. 1" x 6"

25. The BEST way to tell whether a ladder is safe for use is to 25.____

A. place it horizontally on two supports and jump on it
B. inspect it visually
C. place it vertically and carefully place heavy weights on it
D. place it vertically and have several men climb it at the same time

26. The one of the following which should NOT be used on a wooden ladder is: 26.____

A. paint
B. varnish
C. shellac
D. linseed oil

27. Wooden ladders stored horizontally should be supported 27.____

A. throughout their length (that is, resting on the floor)
B. by blocking at each end
C. by two blocks near the middle
D. ~~by blocking at the ends and at intermediate points~~

28. A safety (life) line is *normally* used with a 28.____

A. ladder
B. swinging scaffold
C. single pole scaffold
D. step ladder

29. The hook on the lower pulley block of a swinging scaffold *should* 29.____

A. be moused with marlin
B. be moused with wire
C. NOT be moused
D. be free so that it can be unhooked instantly

30. Nylon rope is better suited for life lines than manila rope *primarily* because of its 30.____

A. elastic quality
B. strength
C. smoothness
D. freedom from decay

31. Manila rope should be stored in a 31.____

A. cool, dry, well-ventilated room
B. cool, dry room
C. warm, dry, well-ventilated room
D. cool, damp room

32. Cracked and scaled paint on a wooden wall can BEST be removed by 32.____

A. scraping
B. wire brushing and scraping
C. burning and scraping
D. burning

33. When sanding wood, the sandpaper should be moved 33.____

A. at right angles to the grain
B. diagonally across the grain
C. parallel to the grain
D. in a circular path

34. In comparing the lasting quality of a paint job on exterior wood doors done with three coats of a high grade oil paint as compared to three coats of spar varnish, the oil paint job will last _______ time. 34.____

A. about the same length of
B. a much longer
C. a much shorter
D. only a little longer

35. The total thickness of a three-coat paint job on exterior woodwork *should be* approximately _______ a two-coat paint job. 35.____

A. twice the thickness of
B. 50% thicker than the thickness of
C. 25% thicker than the thickness of
D. the same thickness as

KEY (CORRECT ANSWERS)

1.	D	16.	A
2.	B	17.	B
3.	C	18.	B
4.	D	19.	B
5.	D	20.	B
6.	B	21.	C
7.	A	22.	C
8.	B	23.	C
9.	C	24.	D
10.	D	25.	B
11.	D	26.	A
12.	B	27.	D
13.	A	28.	B
14.	A	29.	C
15.	C	30.	A

31.	A
32.	C
33.	C
34.	B
35.	D

EXAMINATION SECTION
TEST 1

DIRECTIONS: Each question or incomplete statement is followed by several suggested answers or completions. Select the one that BEST answers the question or completes the statement. *PRINT THE LETTER OF THE CORRECT ANSWER IN THE SPACE AT THE RIGHT.*

1. When painting raw wood, puttying of nail holes should be done 1.____

   A. 24 hours before the prime coat
   B. immediately before the prime coat
   C. after the prime coat and before the second coat
   D. after the second coat and before the finish

2. In general, the one of the following that will dry *tack-free* in the SHORTEST time is 2.____

   A. lacquer B. varnish C. enamel D. oil paint

3. The *vehicle* MOST frequently used in paints for exterior wood surfaces is 3.____

   A. white lead B. linseed oil
   C. japan D. varnish

4. Painting of an interior plastered wall is usually delayed until the plaster is dry. If this practice is NOT followed, the paint might 4.____

   A. chalk B. fade C. run D. blister

5. A *sealer* applied over knots and pitch streaks to prevent *bleeding* through paint is 5.____

   A. shellac B. lacquer
   C. coal tar D. carnauba wax

6. Painting of outside steel in near freezing (32°F) weather is poor practice MAINLY because 6.____

   A. the paint will not dry properly
   B. ice will form in the thinner
   C. more paint is required
   D. paint fumes are dangerous

7. When repainting exterior woodwork that has a glossy finish, good adhesion of paint is BEST obtained by first 7.____

   A. washing the work with diluted lye
   B. dulling the work with sandpaper
   C. warming the work with an electric heater
   D. roughening the work with a rasp

8. The one of the following methods of cleaning steelwork prior to painting that is NOT commonly used on exterior work, such as bridges, is 8.____

   A. sandblasting B. flame cleaning
   C. wire brushing D. pickling

9. When spraying oil paints, the type of gun and nozzle preferred is a _______ feed gun, _______ mix nozzle. 9._____

A. pressure; internal
B. pressure; external
C. syphon; internal
D. syphon; external

10. Red lead paint would MOST likely be used 10._____

A. as a prime coat on steel surfaces
B. as a finish coat on plastered walls
C. as a base coat for concrete surfaces
D. where a glossy finish is desired

11. A permit to store paint in quantities greater than 20 gallons must be obtained from the 11._____

A. Police Department
B. EPA
C. Fire Department
D. Building Department

12. Before painting a kitchen wall, 12._____

A. a degreaser must be mixed with the paint
B. all traces of grease must be washed off
C. a water-base paint must be used to dissolve the grease
D. the walls must be sanded to remove all traces of grease and old paint

13. For interior walls which must be washed very often, the PREFERRED paint is 13._____

A. enamel
B. flat
C. exterior varnish
D. calsomine

Questions 14-15.

DIRECTIONS: Questions 14 and 15 are to be answered in accordance with the following paragraph.

Painting is done to preserve surfaces; and unless the surface is properly prepared, good preservation will not be possible. Apply paint only to clean, dry surfaces. After a surface has been scaled, which means that all loose paint and rust are removed by chipping, scraping, and wire brushing, be sure all dust and dirt are completely removed.

14. According to the above paragraph, the MAIN purpose of painting a wall is to _______ the wall. 14._____

A. clean
B. waterproof
C. protect
D. remove dust from

15. According to the above paragraph, 15._____

A. chipping, scraping, and wire brushing are the only methods permitted for cleaning surfaces
B. painting is effective only when the surface is clean
C. scaling refers only to the removal of rust
D. paint may be applied on wet surfaces

16. The ceiling of a room which measures 20' x 30' is to be given two coats of paint. If one gallon of paint will cover 500 square feet, the two coats of paint will require a MINIMUM of _______ gallons. 16._____

A. 1.5
B. 2
C. 2.4
D. 3.2

17. The type of paint that uses water as a thinner is 17.____

A. enamel B. latex C. shellac D. lacquer

18. The PROPER cleaning agent for a paint brush that has been used to shellac a floor is 18.____

A. gasoline
B. linseed oil
C. alcohol
D. turpentine

19. Under the same conditions, the one of the following that dries the FASTEST is 19.____

A. shellac B. varnish C. enamel D. lacquer

20. When painting walls with two coats of paint, a different color is used for each coat PRIMARILY to 20.____

A. check for full coverage by the second coat
B. provide a better appearance
C. lower the painting cost
D. allow the painter to use any color paint for the first coat

21. Which one of the following statements is NOT correct? 21.____

A. Boiled linseed oil is used for painting wood exposed to the weather.
B. Turpentine is added to final paint coats when a dull finish is desired.
C. A priming coat containing turpentine will penetrate better than one with no turpentine.
D. If too much drier is used, a paint is likely to crack and disintegrate.

22. When painting wood, puttying of nail holes and cracks is done 22.____

A. before any painting is started
B. after the priming coat is applied
C. after the finish coat is applied
D. at any stage in the painting

23. The process of pouring paint from one container to another in order to mix it is known as 23.____

A. bleeding B. boxing C. cutting D. stirring

24. Before repainting a wood surface on which the old paint film has developed some wrinkling, the MOST appropriate treatment for the wood surface is a 24.____

A. thorough scraping
B. light shellacking
C. wash-down with dilute muriatic acid
D. rubbing down of the wrinkles with fairly coarse sandpaper

25. When a common straight ladder is used to paint a wall, the safe distance that the foot of the ladder should be set away from the wall is MOST NEARLY ______ the length of the ladder. 25.____

A. one-eighth
B. one-quarter
C. one-half
D. five-eighths

KEY (CORRECT ANSWERS)

1. C
2. A
3. B
4. D
5. A
6. A
7. B
8. D
9. A
10. A
11. C
12. B
13. A
14. C
15. B
16. C
17. B
18. C
19. D
20. A
21. A
22. B
23. B
24. D
25. B

TEST 2

DIRECTIONS: Each question or incomplete statement is followed by several suggested answers or completions. Select the one that BEST answers the question or completes the statement. *PRINT THE LETTER OF THE CORRECT ANSWER IN THE SPACE AT THE RIGHT.*

1. The one of the following that is NOT a defect in painting is 1.____

   A. chalking
   B. checking
   C. alligatoring
   D. waning

2. The one of the following ingredients of a paint that would be called the *vehicle* is 2.____

   A. white lead
   B. turpentine
   C. linseed oil
   D. pigment

3. The one of the following that is used as a rust preventative in the prime coat for painting steel is 3.____

   A. aluminum
   B. red lead
   C. titanium dioxide
   D. carbon black

4. *Boxing,* with reference to paint, means 4.____

   A. thinning B. mixing C. spreading D. drying

5. Assume that a shop is undergoing a general housecleaning, and all excess unused materials have been removed. 5.____
   Clean-up work, as pertains to painting in this case, means MOST NEARLY

   A. a thorough two-coat paint job
   B. only that surface which was marred to be painted
   C. a one-coat job to *freshen things up*
   D. only that iron work is to be painted

6. A room is 7'6" wide by 9'0" long with a ceiling height of 8'0". One gallon of flat paint will cover approximately 400 square feet of wall. 6.____
   The number of gallons of this paint required to paint the walls of this room, making no deductions for windows or doors, is MOST NEARLY _____ gallon.

   A. 1/4 B. 1/2 C. 3/4 D. 1

Questions 7-8.

DIRECTIONS: Questions 7 and 8 are to be answered on the basis of the following statement.

Surfaces of woodwork shall be in proper condition by sanding all edges smooth to receive prime coat. The paint shall be uniformly applied and, if by brush, well-brushed into all cracks and crevices. Undercoat shall be well-sanded before application of final enamel.

7. Of the following items, the one to which this statement MOST likely applies is 7.____

   A. finished hardware
   B. overhead clothes dryers
   C. baseboards
   D. kitchen cabinets

8. According to the above statement, application of the paint by brush is 8.____

A. optional
B. preferable
C. essential
D. required if there are cracks in the wood

9. Those materials which are added to a paint vehicle to regulate its consistency and thus increase its spreading power and facilitate its application are called 9.____

A. driers B. thinners C. extenders D. oxidents

10. A COMMON example of a paint thinner is usually 10.____

A. tung oil
B. chinawood oil
C. lead oxide
D. turpentine

11. In the painting of rooms in a housing project or school by the contractor, the superintendent representing the city is LEAST concerned with 11.____

A. the area covered per man per day
B. whether the paint is being used at the required spreading rate
C. the moisture content of the plaster
D. the condition of the surfaces to be painted

12. From experience, you have found that one gallon of primer will cover 600 square feet and one gallon of interior latex flat paint will cover 400 square feet. 12.____
If you estimate that a certain repair job has 3,600 square feet to be painted and that both a coat of primer and a coat of flat paint are required, the material required for this job is ______ gallons of primer and ______ gallons of flat paint.

A. 6; 6 B. 6; 9 C. 3; 9 D. 9; 6

13. Crocus cloth is COMMONLY used to 13.____

A. protect finely machined surfaces from damage while the machines are being repaired
B. remove rust from steel
C. protect floors and furniture while painting walls
D. wipe up oil and grease that has spilled

14. Before using a new paint brush, the FIRST operation should be to 14.____

A. remove loose bristles
B. soak the brush in linseed oil
C. hang the brush up overnight
D. clean the brush with turpentine

15. The vehicle used with latex paints is USUALLY 15.____

A. linseed oil
B. shellac
C. varnish
D. water

16. Assume that you are to make an inspection of a building to determine the need for painting. 16.____
Of the following tools, the one which is LEAST needed to aid you in your inspection is a

A. sharp penknife
B. putty knife
C. lightweight tack hammer
D. six-foot rule

17. Assume that a can of red lead paint needs to be thinned slightly. 17.____
Of the following, the one that should be used is

A. turpentine
B. lacquer thinner
C. water
D. alcohol

18. Of the following types of paint, the one that can MOST readily be applied by spraying is 18.____

A. lacquer
B. shellac
C. varnish
D. bituminous-based paints

19. The PRIMARY reason for painting is to _______ a surface. 19.____

A. beautify
B. protect
C. hide
D. change the color of

20. Which of the following will have the LEAST detrimental effect on the painting of a surface? 20.____

A. Moisture
B. Dust
C. Grease
D. 80° heat

21. Of the following, the finish that would MOST likely be used on a hardwood floor is 21.____

A. enamel paint
B. wood stain
C. gutta percha
D. shellac

22. Because of its weather-resistant properties, a varnish COMMONLY used on exterior surfaces is _______ varnish. 22.____

A. spar
B. flat
C. rubbing
D. hard oil

23. Of the following, the BEST treatment for knots in woodwork that is to be painted is to coat the knots with 23.____

A. shellac
B. colorless stain
C. boiled linseed oil
D. raw linseed oil

24. A brush that has been used in shellac should be cleaned by washing it in 24.____

A. water
B. linseed oil
C. lacquer thinner
D. alcohol

25. Excessive moisture on a surface being painted would MOST likely result in 25.____

A. alligatoring
B. blistering
C. cracking
D. sagging

KEY (CORRECT ANSWERS)

1.	D	11.	A
2.	C	12.	B
3.	B	13.	B
4.	B	14.	A
5.	C	15.	D
6.	C	16.	D
7.	D	17.	A
8.	A	18.	A
9.	B	19.	B
10.	D	20.	D

21.	D
22.	A
23.	A
24.	D
25.	B

TEST 3

DIRECTIONS: Each question or incomplete statement is followed by several suggested answers or completions. Select the one that BEST answers the question or completes the statement. *PRINT THE LETTER OF THE CORRECT ANSWER IN THE SPACE AT THE RIGHT.*

1. The vehicle for an oil paint is USUALLY 1.____

   A. soybean oil
   B. methyl alcohol
   C. linseed oil
   D. turpentine

2. Whitewash is composed PRIMARILY of 2.____

   A. lead oxide and water
   B. gypsum and water
   C. methyl alcohol and gypsum
   D. lime and water

3. In preparing a wood surface prior to painting, knots should USUALLY be 3.____

   A. removed and the space filled with putty
   B. coated with shellac
   C. sandpapered to a fine finish
   D. given no special treatment

4. Enamel is GENERALLY paint 4.____

   A. containing lithopone as a pigment
   B. having little hiding power
   C. containing organic pigments
   D. containing varnish as a vehicle

5. Of the following, the BEST reason for using wood stain on woodwork is that it 5.____

   A. bleaches the wood surface
   B. covers the defects of the wood
   C. brings out the texture of the wood
   D. etches the wood surface

6. A painting specification for a large job may require a different tint for each of three coats. The reason for this requirement is MOST likely that 6.____

   A. it enables the painter to get the desired shade in the final coat
   B. it is impossible to obtain the same tint in the types of paint required for the three different coats
   C. it reduces the cost of the job
   D. if a painter skimps, the undercoat will show through

7. New plaster walls that are to be painted with an oil base paint 7.____

   A. require no sealer
   B. should be sealed with a coat of shellac
   C. should first receive a coat of primer sealer
   D. should first be brushed with turpentine

8. Of the following, the BEST thinner for an oil base paint is 8.____

A. turpentine
B. benzine
C. alcohol
D. glycerine

9. *Cutting in* is done when 9.____

A. trimming a stud to size
B. fitting a bat in a brick wall
C. painting in tight corners
D. trimming tallow for a wiped joint

10. In painting new concrete masonry surfaces, it is BEST that such surfaces first be 10.____

A. dry-brushed
B. white-washed
C. wet down
D. oiled

11. Red lead is often used as a pigment in metal priming paints PRIMARILY because it 11.____

A. provides good coverage
B. presents a good appearance
C. makes painting easier
D. is a rust inhibitor

12. The weight of a gallon of ordinary paint, in pounds, is MOST NEARLY 12.____

A. 5 B. 13 C. 21 D. 29

13. The coloring material in an exterior wall paint is called the 13.____

A. solvent B. lacquer C. vehicle D. pigment

14. A room 20' x 25' in area with a ceiling height of 9'6" is to be painted. One gallon of paint will cover 400 square feet. 14.____
The MINIMUM number of gallons necessary to give the four walls and the ceiling one coat of paint is

A. 2 B. 3 C. 4 D. 5

15. *Boxing* is done in order to ensure proper ______ of paint. 15.____

A. consistency
B. air content
C. temperature
D. volume

16. The object LEAST likely to be spray-painted is a(n) 16.____

A. nursery room
B. chain-link fence
C. exterior wall
D. garage door

17. A paint that has a dull finish is called a(n) ______ paint. 17.____

A. gloss B. enamel C. flat D. chalk

18. Whitewash is MOST likely to be used on a(n) 18.____

A. bedroom wall
B. kitchen cabinet
C. picket fence
D. basement floor

19. Latex paint would LEAST likely be used on ______ walls. 19.____

A. exterior B. bedroom C. office D. playroom

20. Enamel paints USUALLY dry with a(n) ______ finish. 20.____

A. glossy B. marbled C. flat D. tacky

KEY (CORRECT ANSWERS)

1.	C	11.	D
2.	D	12.	B
3.	B	13.	D
4.	D	14.	C
5.	C	15.	A
6.	D	16.	B
7.	C	17.	C
8.	A	18.	C
9.	C	19.	A
10.	A	20.	A

EXAMINATION SECTION
TEST 1

DIRECTIONS: Each question or incomplete statement is followed by several suggested answers or completions. Select the one that BEST answers the question or completes the statement. *PRINT THE LETTER OF THE CORRECT ANSWER IN THE SPACE AT THE RIGHT.*

Questions 1-28.

DIRECTIONS: Questions 1 through 28 are concerned with terms commonly used in the painting trade. Answer these questions by selecting the meaning as used in the trade.

1. BOXING is 1.____
   A. brushing B. mixing C. enclosing D. scaling

2. A HOLIDAY is a ______ spot. 2.____
   A. hlgh B. faded C. depressed D. skipped

3. FLAG ends are found on 3.____
   A. ropes B. rollers C. ladders D. brushes

4. FOOTS refer to 4.____
   A. sediment B. skin C. crust D. bubbles

5. BOILED OIL is ______ oil. 5.____
   A. fish B. linseed C. rung D. cottonseed

6. WHIPPING is used with 6.____
   A. rollers B. brushes C. ropes D. ladders

7. JAPAN is a(n) 7.____
   A. pigment B. drier C. thinner D. extender

8. CALCIMINE is a(n) ______ paint. 8.____
   A. water B. latex C. oil D. alkyd

9. GRAPHITE is colored 9.____
   A. white B. black C. gold D. silver

10. CHINAWOOD oil is ______ oil 10.____
    A. tung B. linseed
    C. cottonseed D. soybean

11. LITHOPONE contains 11.____
    A. lead carbonate B. lead sulfate
    C. zinc sulfide D. zinc oxide

12. TURPENTINE is made from 12.___

A. resin of pine trees
B. benzine
C. coal tar
D. petroleum

13. PICKLING is 13.___

A. drying
B. preserving
C. cleaning
D. coloring

14. BLISTERING is caused by 14.___

A. overthinning
B. overheating
C. moisture
D. low temperature

15. EXTENDERS are 15.___

A. driers
B. inert fillers
C. stainers
D. vehicles

16. The VEHICLE in a paint is the 16.___

A. liquid B. base C. body D. pigment

17. RAW SIENNA is colored 17.___

A. black B. yellow C. red D. white

18. TITANIUM OXIDE is a 18.___

A. solvent B. drier C. pigment D. extender

19. LINSEED OIL is made from 19.___

A. hempseed
B. cottonseed
C. poppy seed
D. flaxseed

20. BURNT UMBER is colored 20.___

A. yellow B. white C. red D. brown

21. PIGMENT particles are bound together by 21.___

A. driers B. vehicles C. solvents D. extenders

22. CHECKING is 22.___

A. bubbling
B. blistering
C. shrinking
D. expansion

23. The COBALT used in tinting is 23.___

A. yellow B. blue C. brown D. red

24. CARNAUBA is a 24.___

A. wax B. pigment C. putty D. plaster

25. The MAIN disadvantage of WHITE LEAD was that it was 25.___

A. difficult to tint
B. poisonous
C. expensive
D. hard to thin

26. EFFLORESCENCE is normally found on walls made of 26.____

A. cinder block
B. gypsum block
C. plaster
D. brick

27. MILDEW is usually caused by 27.____

A. high temperature
B. low temperature
C. insufficient mixing
D. moisture

28. The main purpose of a SOLVENT is to 28.____

A. make it easier to brush out the paint
B. lower the cost of the paint
C. strengthen the paint film
D. increase the life of the paint

29. The purpose of adding zinc oxide to the formula for an exterior white house paint is to make the resulting paint film 29.____

A. whiter
B. harder
C. smoother
D. softer

30. A paint of good quality wears away by 30.____

A. scaling
B. chalking
C. cracking
D. flaking

31. Before a new hog-bristle paint brush is used for the first time with an oil paint, it would be good practice to soak the brush in 31.____

A. benzine for eight hours
B. water for two hours
C. linseed oil overnight
D. carbon tetrachloride overnight

32. The BEST type of brush to use for housepainting is one that is made of 32.____

A. synthetic bristle
B. nylon
C. horsehair
D. boar bristle

33. The area where manila scaffold ropes are stored should be 33.____

A. cool and dry
B. cool and humid
C. hot and dry
D. hot and humid

34. Application of an excessively heavy coat of paint which has too great an oil content will MOST likely cause 34.____

A. washing
B. wrinkling
C. blistering
D. alligatoring

35. Scaffold fibre ropes which have been exposed to muriatic acid should be 35.____

A. discarded immediately
B. treated with sodium hydroxide
C. oiled with warm linseed oil
D. washed with soap and warm water

KEY (CORRECT ANSWERS)

1. B
2. D
3. D
4. A
5. B

6. C
7. B
8. A
9. B
10. A

11. C
12. A
13. C
14. C
15. B

16. A
17. B
18. C
19. D
20. D

21. B
22. C
23. B
24. A
25. B

26. D
27. D
28. A
29. B
30. B

31. C
32. D
33. A
34. B
35. A

TEST 2

DIRECTIONS: Each question or incomplete statement is followed by several suggested answers or completions. Select the one that BEST answers the question or completes the statement. *PRINT THE LETTER OF THE CORRECT ANSWER IN THE SPACE AT THE RIGHT.*

1. Before a rusty iron surface is painted, it should be 1.____

   A. wire brushed
   B. treated with shellac
   C. washed with kerosene
   D. varnished

2. If it is necessary to paint a new concrete floor that has not had time to dry thoroughly, it should be treated with a solution of 2.____

   A. zinc sulphate
   B. sodium chloride
   C. soap and water
   D. hydrochloric acid

3. The main reason why it is NOT considered good practice to paint the portable ladders used by painters is that the paint 3.____

   A. will quickly wear off
   B. may mar the work area
   C. may hide serious defects
   D. may make the ladder slippery

4. Painting specifications for a large job frequently require a different tint for each of the three coats. 4.____
   The MAIN reason for this requirement is that it

   A. helps spot missed areas
   B. reduces cost since only one paint need be purchased
   C. helps the painter to get the exact shade for the final coat
   D. is not feasible to get the right shade for all coats

5. Paint mildew is BEST removed by 5.____

   A. applying paint with a high oil content
   B. covering it with a slow-drying paint
   C. washing it with linseed oil
   D. washing it with warm water and strong soap

6. The color which should be added to a yellow paint in order to *gray* it is 6.____

   A. orange
   B. red
   C. violet
   D. green

7. A metal paint that is to be sprayed differs from one that is to be brush-applied in that it 7.____

   A. contains more oil
   B. contains more drier
   C. contains more pigment
   D. is of a thinner consistency

8. Of the following, the BEST method of removing all foreign matter from structural steel prior to painting is 8.____

A. sandblasting
B. wire brushing
C. scraping
D. chipping

9. If the specifications call for the *stringer* to be enameled, the painter should check the paint on the 9.____

A. roof beams
B. door frames
C. stairs
D. windows

10. The PRIMARY objective in drawing up a set of specifications for painting materials is the 10.____

A. control of quality
B. outlining of intended use
C. establishment of standard coverage
D. establishment of inspection procedures

11. The varnish that is USUALLY specified because of its weather-resistant properties is ______ varnish. 11.____

A. hard oil
B. rubbing
C. flat
D. spar

12. If a cement-block wall is to be painted with a cement-water paint, its surface should be 12.____

A. varnished
B. shellacked
C. dry
D. wetted

13. The BEST way to insure that a paint job on structural steel is long-lasting is to make sure the coats of paint are 13.____

A. rigid and thin
B. rigid and thick
C. elastic and thin
D. elastic and thick

14. The BEST method of removing loose and scaling paint from galvanized ducts that are to be repainted is to use 14.____

A. a hammer and chisel
B. coarse steel wool
C. a putty knife
D. a heavy scraper

15. The BEST material for cleaning brushes that have been used to apply shellac is 15.____

A. soapy water
B. benzine
C. alcohol
D. lacquer thinner

16. The ADVANTAGE of nylon bristle over hog bristle for brushes is that the nylon bristle 16.____

A. will outwear the hog bristle
B. is soluble in phenol
C. does not soften in shellac
D. has a greater pick-up capacity

17. The BEST practice for removing the excess paint on a brush is to 17.____

A. tap the handle on the edge of the paint can
B. tap the bristles lightly against the inside of the paint can

C. rub the bristles lightly across the inside edge of the paint can
D. hold the brush above the paint surface so the excess paint can drip back into the can

18. If paint is applied without being properly mixed, the USUAL result is 18.____

A. cracking B. scaling C. running D. spotting

19. To prevent paint from *crawling* when repainting a glossy surface, the surface should FIRST be 19.____

A. shellacked B. sandpapered
C. washed with naphtha D. wet down with water

20. The BEST thinner for varnishes is 20.____

A. turpentine B. varnolene
C. kerosene D. benzine

21. Before a water-based paint is applied to a recently plastered wall, the wall should be 21.____

A. washed B. shellacked
C. varnished D. sized

22. When painting a steel suspension bridge, a painter should use a *swab* to 22.____

A. clean the steel
B. remove rust
C. spread the paint on rough surfaces
D. get paint into narrow spaces

23. The MAIN purpose of a *drier* is to 23.____

A. prevent the paint from scaling
B. promote the oxidation of the linseed oil
C. increase the penetration of the paint
D. hasten the evaporation of the turpentine

24. A *fugitive* color is one that 24.____

A. has little brightness value
B. fuses easily with other colors
C. has a tendency to run on metal
D. fades in strong light

25. *Spotting* of paint on new wood occurs MAINLY when 25.____

A. industrial gases are present
B. the finish coat lacks elasticity
C. an improper brushing procedure is used
D. an inadequate amount of paint is used

26. *Benzol* should be added to the primary coat applied on 26.____

A. white pine B. chestnut
C. oak D. cypress

27. *Enamel* paint is a paint that contains 27.____

A. lithopone as a pigment
B. organic pigments only
C. varnish as a vehicle
D. whiting as a vehicle

28. The MAIN purpose of having a drying oil in paint is to 28.____

A. permit air to reach the wood surface
B. bind the pigment particles together
C. dull the finish
D. remove moisture from the wood

29. The BEST method of preparing a galvanized surface for painting is to FIRST 29.____

A. prime it with graphite
B. prime it with red lead
C. wet it with a weak solution of muriatic acid
D. wet it with a weak solution of tri-sodium phosphate

30. Before painting an old plastered wall that has been calcimined, it is advisable to FIRST 30.____

A. apply paint remover to the plastered wall
B. apply a solution of muriatic acid to the plastered wall
C. wash off the calcimine with water
D. wash off the calcimine with tri-sodium phosphate

31. A room will appear LARGER if it has been painted with a 31.____

A. dark colored paint on all walls
B. light colored paint on all walls
C. design on at least one wall
D. dark color on two walls, light color on the other walls

32. In sanding a wood surface with steel wool, the stroke should be 32.____

A. with the grain
B. crosswise to the grain
C. circular
D. oblique

33. Small cuts or injuries received on the job should be 33.____

A. ignored since they are minor
B. ignored unless they slow you down
C. taken care of immediately to avoid infection
D. taken care of if they continue to bleed

34. Artificial respiration should be started IMMEDIATELY on a painter who has suffered an electric shock if he is 34.____

A. *unconscious* and badly burned
B. *conscious* and in a daze
C. *unconscious* and breathing heavily
D. *unconscious* and not breathing

35. The MAIN purpose of giving first aid instruction to some selected painters is to 35.____

A. avoid calling a doctor to the job
B. save money
C. enable them to give first aid in an emergency
D. reduce the number of accidents

KEY (CORRECT ANSWERS)

1. A
2. A
3. C
4. A
5. D

6. C
7. D
8. A
9. C
10. A

11. D
12. D
13. C
14. B
15. C

16. A
17. B
18. C
19. B
20. A

21. D
22. D
23. B
24. D
25. D

26. D
27. C
28. B
29. C
30. C

31. B
32. A
33. C
34. D
35. C

EXAMINATION SECTION
TEST 1

DIRECTIONS: Each question or incomplete statement is followed by several suggested answers or completions. Select the one that BEST answers the question or completes the statement. *PRINT THE LETTER OF THE CORRECT ANSWER IN THE SPACE AT THE RIGHT.*

1. Certain specifications require that each coat of paint shall be done with a different tint from the preceding coat. 1.____
Of the following, the BEST reason for this requirement is to

A. insure that the final coat will have a uniform color
B. insure that matching tints give a finer final color
C. enable the inspector to check the number of coats being applied
D. compensate for the loss of color with succeeding coats

2. The BEST way to insure that the finish coat of paint is the exact shade and texture required is to 2.____

A. check the ingredients of the paint
B. check the finish coat against an approved sample surface
C. be sure that the finish coat is pleasing to the eye
D. check the thickness of the coat of paint

3. *Powdering* of the exposed surface of a paint film is known as 3.____

A. sagging B. holidays C. chalking D. running

4. The one of the following which is a characteristic of a lacquer is that it 4.____

A. forms a soft film
B. is odorless
C. is quick drying
D. is difficult to spray on

5. Efflorescence would MOST likely occur on a ______ surface. 5.____

A. brick B. limestone C. granite D. concrete

6. A shellac which is designated as a *4 1/2 lbs. cut shellac* contains 4 1/2 lbs. of resin per ______ of solvent. 6.____

A. 100 lbs. B. pint C. quart D. gallon

7. A reasonable amount of chalking of paint on an exterior surface is desirable because it 7.____

A. permits self-cleaning of the coating
B. increases the thickness of the paint coat
C. reduces glare on the paint surface
D. minimizes the tendency of the paint to blister

8. The MOST immediate effect of leaving a can of paint uncovered for a long time is that the paint will 8.____

A. fade in color
B. ~~become thicker~~
C. lose its hiding power
D. lose its ability to adhere to the surface being painted

9. Leafing would MOST likely occur when using a(n) ______ base paint. 9.____

A. lead B. aluminum C. zinc D. copper

10. The color of a paint using zinc chromate as the only coloring pigment is usually 10.____

A. brown B. red C. yellow D. blue

11. The color of a paint using zinc oxide as the only coloring pigment is 11.____

A. white B. yellow C. green D. blue

12. The PRIMARY colors are 12.____

A. yellow, red, and violet
B. red, yellow, and blue
C. pink, yellow, and green
D. purple, orange, and green

13. Pouring of paint back and forth from one container to another until the pigment and liquid form a smooth mixture of uniform consistency and color is known as 13.____

A. skinning B. milking C. boxing D. canning

14. Of the following, the one that is LEAST likely to be used to remove lumps, skins, and foreign material from a can of paint is 14.____

A. fine wire mesh
B. silk
C. cheesecloth
D. burlap

15. The colors MOST likely to fade in sunlight are 15.____

A. black and brown
B. white and yellow
C. red and orange
D. blue and green

16. The ability of a paint to obscure the underlying surface is known as 16.____

A. range
B. hiding power
C. reflectance
D. solvency

17. Of the following, the chemical that should be used to clean a masonry surface is a dilute solution of ______ acid. 17.____

A. sulphuric
B. hydrochloric
C. hydrofluoric
D. phosphoric

18. The MOST frequent cause of failure of paint on concrete, plaster, and masonry surfaces is 18.____

A. oil on the surface before painting
B. ~~hydrogen sulfide in the air~~
C. carbon dioxide in the air
D. moisture

19. A shop coat of paint on steel is usually applied 19.____

A. *before* the steel is shipped to the job site
B. *after* the steel reaches the job site but before the material is set in place
C. *after* the steel is set in place but before the final coat
D. *just before* the steel is to be encased in concrete

20. Brushes used to apply latex based paints should be cleaned by washing in 20.____

A. alcohol
B. turpentine
C. water
D. naphtha

21. When dipping a brush into paint for painting, the brush should be dipped MOST NEARLY ______ length of the bristles. 21.____

A. 1/4 to 1/3 the
B. 1/3 to 1/2 the
C. 1/2 to 3/4 the
D. 3/4 to the full

22. A room has dimensions of 12' x 18' and is 8 feet high. 22.____
Neglecting door and window openings, the wall area of this room is, in square feet, MOST NEARLY

A. 390
B. 420
C. 450
D. 480

23. The BEST weather condition for outdoor painting of wood surfaces is 23.____

A. damp and cool
B. hot and dry
C. cold and dry
D. warm and clear

24. Red lead would MOST likely be used to prime coat 24.____

A. wood
B. metal
C. plaster
D. concrete

25. For estimating purposes, the FIRST coat coverage of a gallon of white flat wall paint of a three-coat paint job is MOST NEARLY ______ square feet. 25.____

A. 100
B. 250
C. 500
D. 700

KEY (CORRECT ANSWERS)

1. C
2. B
3. C
4. C
5. A
6. D
7. A
8. B
9. B
10. C
11. A
12. B
13. C
14. D
15. D
16. B
17. B
18. D
19. A
20. C
21. B
22. D
23. D
24. B
25. C

TEST 2

DIRECTIONS: Each question or incomplete statement is followed by several suggested answers or completions. Select the one that BEST answers the question or completes the statement. *PRINT THE LETTER OF THE CORRECT ANSWER IN THE SPACE AT THE RIGHT.*

1. Certain specifications state that the top and bottom edges of exterior wood doors immediately after being fitted shall be painted with a heavy coat of enamel undercoat. Of the following, the BEST reason for this requirement is to 1.____

   A. prevent splitting of the top and bottom of the door
   B. conceal the wood joints at the top and bottom
   C. enable the carpenter to tell if the door fits properly
   D. protect the door against the weather

2. A shaking type mixer was used to mix an enamel paint, causing it to bubble. The paint should NOT be used for AT LEAST ______ hour(s). 2.____

   A. 1 B. 2 C. 4 D. 6

3. The type of oil varnish MOST often used for interior painting is ______-oil interior varnish. 3.____

   A. long B. medium
   C. short D. extra short

4. Before applying a second coat of paint, the FIRST coat should be 4.____

   A. wet B. dry only on the surface of the coat
   C. tacky D. thoroughly dry

5. When the operating air pressure of a spray gun is about 30 pounds per square inch, the proper distance to hold the gun's nozzle from the work surface is MOST NEARLY 5.____

   A. 2" B. 7" C. 13" D. 19"

6. Drying time between coats of a varnish base paint should be at least equal to or more than ______ hours. 6.____

   A. 24 B. 72 C. 96 D. 108

7. When painting paneled doors with a brush, you should start at the 7.____

   A. top panel, and do the molding edges first
   B. top rail, and do the top rail first
   C. stile
   D. bottom rail, and do the bottom rail first

8. The one of the following that is a CORRECT practice to use when rolling paint on a ceiling is to 8.____

   A. start at the center of the ceiling
   B. spin the roller frequently
   C. finish with heavy strokes in the cross direction
   D. cross roll from unpainted areas to newly painted sections

9. Areas of a surface that are missed by the painter are known as 9.____

A. voids B. ghosts C. foots D. holidays

10. Ready-mixed paints for immediate use are PREFERRED to paints made up manually by mixing the individual ingredients together mainly because 10.____

A. the individual ingredients in paint are unobtainable
B. ready-mixed paints are more uniform in color and texture
C. manually made-up paints generally contain inferior ingredients compared to ready-mixed paints
D. ready-mixed paints do not have to be stirred before being used

11. Of the following, paint runs on a wall are MOST often caused by 11.____

A. holding a spray gun nozzle too close
B. using the wrong brush
C. using the wrong roller
D. overbrushing the painted surface

12. Blistering of paint on wood surfaces is due mainly to 12.____

A. moisture in the wood
B. rough wood grain
C. knots in the wood
D. inadequate puttying of the wood surface

13. Of the following grades of sandpaper, the COARSEST is No. 13.____

A. 3 B. 2 C. 1 D. 1/0

14. Sanding of new exterior woodwork before painting is 14.____

A. generally unnecessary
B. generally done with #5 sandpaper
C. done only after the prime coat of paint is applied
D. done to remove grease and oil on the unfinished wood surface

15. When nailholes in woodwork are filled with putty prior to painting, the finished surface of the putty should be 15.____

A. below the surface of the wood to allow for expansion of the putty
B. above the surface of the wood
C. flush with the surface of the wood
D. above the wood surface for regular putty and below the surface for glazing compound

16. When *plastic wood* is used to fill nail holes and cracks, the EARLIEST the wood putty should be sanded so as not to cause damage to the putty is ______ after application. 16.____

A. immediately B. one hour
C. one day D. one week

17. When a concrete wall surface is painted with a Portland cement paint, the surface should be 17.____

A. completely dry
B. almost dry
C. thoroughly wet
D. wet or dry, depending on the humidity

18. When a wood surface is painted, knots and pitch streaks should be 18.____

A. removed
B. burnished
C. shellacked
D. caulked

19. Certain specifications state that plaster wall surfaces, except where enamel is used, shall have a final coat of paint stippled with brush or roller to a fine stipple.
Of the following, the MAIN purpose of stippling is to 19.____

A. eliminate holidays
B. remove sags
C. eliminate brush marks
D. prevent fading of the paint

20. The size of sandpaper which should be used to finish new pine and fir flooring is 20.____

A. #0
B. #1 or # 1 1/2
C. #2 or #2 1/2
D. #3 or #4

21. It is PROPER to apply glazing compound to 21.____

A. either wood or metal sash
B. metal sash only
C. wood sash only
D. wood sash if it is a hardwood and to metal sash only if it is steel

22. Most pigmented paints contain mineral fillers known as 22.____

A. extenders
B. thinners
C. absorbers
D. compounders

23. Generally, the percentage of pigment by weight in oil-base house paints is MOST NEARLY 23.____

A. 10-20
B. 30-40
C. 60-70
D. 85-95

24. Of the following, the MAIN purpose of the vehicle in a paint is to 24.____

A. form the film
B. provide the bulk of the weight of the paint
C. give the paint color
D. keep the pigment from settling in the can

25. The binder in a paint 25.____

A. is the part of the paint that evaporates
B. gives the paint its color
C. is the non-evaporating part of the vehicle
D. acts as a mineral filler

KEY (CORRECT ANSWERS)

1.	D	11.	A
2.	D	12.	A
3.	D	13.	A
4.	D	14.	A
5.	A	15.	C
6.	A	16.	C
7.	A	17.	C
8.	D	18.	C
9.	D	19.	C
10.	B	20.	B

21.	A
22.	A
23.	C
24.	A
25.	C

TEST 3

DIRECTIONS: Each question or incomplete statement is followed by several suggested answers or completions. Select the one that BEST answers the question or completes the statement. *PRINT THE LETTER OF THE CORRECT ANSWER IN THE SPACE AT THE RIGHT.*

1. Linseed oil, tung oil, and soybean oil are known as 1.____

   A. volatile oils
   B. drying oils
   C. extenders
   D. fillers

2. An example of a synthetic resin is 2.____

   A. linseed oil
   B. refined fish oil
   C. benzine
   D. epoxy

3. Shellac is a solution of refined lac resin in 3.____

   A. boiled linseed oil
   B. refined naphtha
   C. denatured alcohol
   D. unleaded gasoline

4. A chemical that is used as a pigment in paint is 4.____

   A. sodium chloride
   B. carbon dioxide
   C. lead oxide
   D. carbon tetrachloride

5. The thinner usually used in oil-base paints is 5.____

   A. water
   B. turpentine
   C. alcohol
   D. naphtha

6. Putty is usually composed of 6.____

   A. talc and naphtha
   B. Portland cement and tung oil
   C. alcohol and calcium silicate
   D. whiting and linseed oil

7. The two sizes of paint rollers MOST often used to paint rooms are ______ inches and ______ inches. 7.____

   A. 2; 5
   B. 7; 9
   C. 11; 13
   D. 13; 15

8. Synthetic bristle brushes are usually made of 8.____

   A. teflon
   B. orlon
   C. nylon
   D. acetate

9. The BEST size of brush to use for trim work such as sashes, frames, and molding is ______ to ______ inches. 9.____

   A. 1; 1 1/2
   B. 1 1/2; 3
   C. 3 1/2; 4
   D. 4 1/2; 6

10. Certain paint brushes are being used daily with oil-base paints. When not in use, the bristles will keep BEST if they are suspended in 10.____

    A. linseed oil
    B. water
    C. alcohol
    D. acetone

11. A DISADVANTAGE of spray painting as compared with brush painting is 11.____

A. more paint is wasted in spraying than in brushing
B. spray painting takes longer than brush painting
C. spray painting does not produce a uniform color
D. spray painting leaves streaks

12. A wood ladder should NOT be painted mainly because 12.____

A. the ladder will become slippery
B. the paint may hide the Underwriters Laboratory approval
C. the paint may hide defects in the ladder
D. it is a waste of good paint

13. Of the following, the tool LEAST likely to be used to paint the wire mesh of a chainlink fence is a 13.____

A. paint mitt
B. roller
C. brush
D. spray gun

14. A Federal government requirement for safety on construction jobs is that 14.____

A. hard hats be worn
B. safety shoes be worn
C. gloves be worn
D. a first aid station be set up on all jobs

15. The types of spray gun nozzles are 15.____

A. internal mix and external mix
B. high pressure and low pressure
C. narrow spray and wide spray
D. self-cleaning and manual cleaning

16. The PRINCIPAL types of paint rollers are pan type, fountain type, and ______ type. 16.____

A. suction
B. pressure
C. vacuum
D. siphon

17. Lead-base paints should NOT be used to paint rooms because lead-base paints 17.____

A. are expensive
B. could prove harmful to children
C. have low coverage
D. chalk easily

18. In interior painting, good ventilation must be provided to prevent concentration of fumes that 18.____

A. will accelerate drying of the paint
B. may cause blistering of the paint
C. may be harmful to the painter
D. may discolor other finished surfaces

19. If some caustic soda solution gets on the skin, the area of contact should be 19.____

A. kept dry
B. rubbed with vaseline
C. washed with water
D. washed with a lime solution

20. A 16-foot ladder should be placed against a vertical wall with the bottom of the ladder ______ feet from the wall. 20.____

A. 2 B. 4 C. 6 D. 8

21. When held at a slight distance, alternating blue and yellow lines of equal width, drawn close and parallel, will appear as 21.____

A. dark yellow
B. light blue
C. gray
D. blue-green
E. green

22. To subdue a bright color and still retain its harmony in a scheme, it is BEST to add a little of 22.____

A. black
B. an adjacent color
C. white
D. its complement
E. a darker tone of the same color

23. Which of these colors is neutral? 23.____

A. Blue B. Gray C. Green D. Red E. Yellow

24. Red and green, when combined, will produce the same color as 24.____

A. orange, yellow, and violet
B. red, blue, and yellow
C. red, blue, and violet
D. red and black
E. yellow and green

25. The BEST color for the walls of a kitchen with a northern exposure is 25.____

A. blue B. green C. violet D. rose E. yellow

KEY (CORRECT ANSWERS)

1. B
2. D
3. C
4. C
5. B

6. D
7. B
8. C
9. B
10. A

11. A
12. C
13. D
14. A
15. A

16. B
17. B
18. C
19. C
20. B

21. E
22. D
23. B
24. B
25. E

PAINT AND PAINTING

TABLE OF CONTENTS

PAINT AND PAINTING

INTRODUCTION

A good paint job not only adds beauty to your home, both interior and exterior; it also protects your investment. The right paint, properly applied to a surface carefully prepared, is an excellent barrier to weathering and decay.

But therein lies a problem. Surface preparation is an exacting task. The paint must be of good quality, and of a type designed for the particular job you want done. And it must be applied as the manufacturer intended or it may not cover well or dry properly to provide a satisfactory appearance.

This guide is for the non-professional painter. Its simple, clear-cut instructions will help you to do a good job of painting the interior and exterior of your home with a minimum of trouble and expense.

Because there are so many types of modern paints, however, it is essential that you read labels carefully and follow instructions exactly as recommended by the manufacturer. The guidelines in this booklet are general in nature; the instructions from the manufacturer are specific to the particular product you are using. For the best results, use both of them.

CAUTION!

Household paints generally are harmless if ventilation is adequate and direct contact with the skin is brief. But people differ in their tolerance for paint, and certain precautions should be taken.

VAPORS

Brief inhalation of paint vapors is rarely harmful, but excessive inhalation may irritate nasal membranes and cause headache, dizziness, nausea and fatigue. Before you paint, be sure you have plenty of ventilation.

CONTACT WITH SKIN

Remove paint promptly after you are through for the day. Extended contact may cause irritation which could lead to infection.

FIRE

Many paint products are flammable. Water thinned types are not. Avoid smoking and open fires when you are painting in an enclosed area. Clean up promptly after the job is finished, and dispose of cleanup rags. Do not store paint near a furnace or radiator.

PAINT AND CHILDREN

Keep every sort of paint and other paint products out of reach of children. If swallowed, paint, varnish, enamel, thinner or paint remover can be extremely dangerous.

METHODS OF APPLYING PAINT

In order to do a good job with a minimum of trouble, choose the right tools and learn how to handle them properly. The brush, the roller and the sprayer are the basic tools to work with.

Using a Brush

The use of a brush assures good contact of paint with pores, cracks and crevices. Brushing is particularly recommended for applying primer coats and exterior paints.

In selecting a brush you should choose one which is wide enough to cover the area in a reasonable amount of time. If you are painting large areas such as exterior or interior walls or a floor, you will want a wide brush—probably four or five inches in width. If you are painting windows or trim, you will want a narrower brush so that you can handle comparatively narrow surfaces—probably 1 to 1-1/2 inches in width.

The bristles should be reasonably long and thick so that they will hold a good load of paint; and flexible, so that you can stroke evenly and smoothly.

Generally speaking, a medium-priced brush is the best investment if you do only occasional painting jobs.

Paint should be brushed up and down, then across for even distribution. On a rough surface, however, it is wise to vary the direction of the strokes so that the paint will penetrate thoroughly. The brush should be held at a slight angle when applying the paint, and pressure should be moderate and even. Excessive pressure or "stuffing" the brush into corners and cracks may damage the bristles.

Always start painting at the top and move downward. For interior painting, do ceilings and walls first, then the doors, windows and trim areas. If floors are to be painted, they should be last. Always work toward the "wet edge" of the previously painted area, making sure not to try to cover too large a surface with each brushload.

Brush Care

A good brush is an expensive tool, and it pays to invest the necessary time and effort to take care of it properly. Clean brushes immediately after use with a thinner or special brush cleaner recommended by your paint or hardware store. Use turpentine or mineral spirits to remove oil base paints, enamels and varnish; alcohol to remove shellac; and special solvents to remove lacquer. Remove latex paints promptly from brushes with soap and water. If any type paint is allowed to dry on a brush, a paint remover or brush-cleaning solvent will be needed.

How to Clean Brushes:

- After removing excess paint with scraper, soak brush in proper thinner; work it against bottom of container.
- To loosen paint in center of brush, squeeze bristles between thumb and forefinger, then rinse again in thinner. If necessary, work brush in mild soap suds, rinse in clear water.
- Press out water with stick.
- Twirl brush—in a container so you won't get splashed.

- Comb bristles carefully—including those below the surface. Allow the brush to dry by suspending from the handle or by laying it flat on a clean surface. Then wrap the dry brush in the original wrapper or in heavy paper to keep the bristles straight. Store suspended by handle or lying flat.

Using a Roller

For large, flat surfaces, painting by roller is easier than painting by brush for the average do-it-yourself painter. Select a roller with a comfortable-to-hold handle and try several dry sweeps across the surface until you get the hang of it.

When you buy a roller you find that it comes as part of a set—the roller itself and a sloping metal or plastic tray. Pour paint into the tray until approximately two-thirds of the corrugated bottom is covered. Dip the roller into the paint in the shallow section of the tray and roll it back and forth until it is well covered. If the roller drips when you lift it from the tray, it is overloaded. Squeeze out some of the paint by pressing the roller against the upper part of the tray above the paint line.

Apply paint by moving the roller back and forth over the surface being painted, first up and down in long, even strokes, then across. Reload the roller with paint as needed.

Roller Care

Rollers used with alkyd or oil base paints should be cleaned with turpentine or mineral spirits. When latex paint has been used, soap and water will do a satisfactory cleaning job. If any kind of paint has been allowed to dry on the roller, a paint remover or brush-cleaning solvent will be needed.

Using a Sprayer

Paint sprayers are particularly useful for large areas. Spraying is much faster than brushing or rolling and, although some paint will likely be wasted through overspraying, the savings in time and effort may more than compensate for any additional paint cost. Once you have perfected your spraying technique, you can produce a coating with excellent uniformity in thickness and appearance.

Surface areas accessible only with difficulty to the brush or roller can readily be covered by the sprayer. All coats can be applied satisfactorily by the spray technique *except for the primer coats.* Spraying should be done only on a clean surface since the paint may not adhere well if a dust film is present.

Pre-preparation of the paint is of critical importance, however, when a sprayer is to be used. Stir or strain to remove any lumps, and thin carefully. If the paint is lumpy or too thick it may clog the spray valve; if it is too thin the paint may sag or run after it is applied. Follow the manufacturer's instructions on the paint label for the type and amount of thinner to be used.

Before you begin, ask your paint dealer to show you exactly how the sprayer works, and to give you pointers on how to use it to its best advantage.

For best results:

- Adjust the width of the spray fan to the size of the surface to be coated. A narrow fan is best for spraying small or narrow surfaces; a wider fan should be used to spray table tops or walls.
- Before spraying, test the thickness of the paint, the size of the fan and the motion of the spray gun before painting any surface. Excessive thickness can cause rippling of the wet film or lead to blistering later.
- Hold the nozzle about eight inches from the surface to be painted.
- Start the stroke or motion of the hand holding the sprayer while the spray is pointed slightly beyond the surface to be painted. This assures a smooth, even flow when you reach the surface to be coated.
- Move the sprayer parallel to the surface, moving with an even stroke back and forth across the area. Spray corners and edges first.
- Use a respirator to avoid inhalation of vapors.
- Cover everything close to the work area with drop cloths, tarps or newspapers. The "bounceback" from a sprayer may extend several feet from the work surface.

Paint Sprayer Care

Clean sprayer promptly before the paint dries. After using oil base or alkyd paints, clean the sprayer with the same solvent used to thin the paint. After using latex paint, clean with detergent and water. Fill the sprayer tank with the cleaning liquid and spray it clean. If the fluid tip becomes clogged, it can be cleaned with a broom straw. Never use wire or a nail to clear clogged air holes in the sprayer tip.

PREPARATION FOR PAINTING

Before you brush, roll or spray a drop of paint, there are certain preparations you should make to ensure a good job with a minimum of effort, errors and spattering. The precautions may seem obvious, but they are often overlooked.

Protect Other Surfaces

Cover floors and furnishings with drop cloths. You can use tarps, old sheets or the inexpensive plastic sheets designed for the purpose. Clean up as you paint. Wet paint is easy to remove; dry paint is hard to remove. Use turpentine or other thinner to remove oil paint; water to remove latex.

CAUTION: *If paint is dropped on an asphalt tile floor, do not attempt to remove it with mineral spirits or turpentine since this may permanently damage the tile. If the paint will not come off with a dry cloth, let it dry and then scrape it off.*

Rub protective cream onto your hands and arms. A film of this cream will make it easier to remove paint from your skin when the job is done. Old gloves or throwaway plastic gloves and aprons are also useful.

Check the Condition of the Paint

When you buy new paint of good quality from a reputable store, it is usually in excellent condition. However, after stirring the paint thoroughly (if it is a type which should be stirred), you should examine it for lumps, curdling or color separation. Do not use the paint if there are still any signs of these conditions.

Old paints which, upon removal of the container lid, release a foul odor (especially latex paints) or show signs of lumps or curdling, are probably spoiled and should be discarded. If there is a "skin" on the surface of the paint when you open the container, remove as much of the hardened film as possible with a spatula or knife and strain the paint through a cheesecloth or fine wire mesh such as window screening. If you fail to do this, bits of the skin will show up with exasperating frequency to spoil the appearance of your paint job.

Follow Directions on Mixing

New paints are usually ready for use when purchased and require no thinning except when they are to be applied with a sprayer. Get the advice of the paint store salesman when you buy the paint, *and check the label before you mix or stir.* Some manufacturers do not recommend mixing as it may introduce air bubbles.

If mixing is required, it can be done at the paint store by placing the can in a mechanical agitator—or you can do it at home with a paddle or spatula. If you open the can and find that the pigment has settled, use a clean paddle or spatula and gradually work the pigment up from the bottom of the can, using a circular stirring motion. Continue until the pigment is thoroughly and evenly distributed, with no signs of color separation.

If the settled layer should prove to be hard or rubbery, and resists stirring, the paint is probably too old and should be discarded.

Protect the Paint Between Jobs

Between jobs, even if it is only overnight, cover the paint container tightly to prevent evaporation and thickening, and to protect it from dust. Oil base and alkyd paints may develop a skin from exposure to the air.

When you finish painting, clean the rim of the paint can thoroughly and put the lid on tight. To ensure that the lid is airtight, cover the rim with a cloth or piece of plastic film (to prevent spattering) and then tap the lid firmly into place with a hammer.

PREPARE THE SURFACE

The finest paint, applied with the greatest skill, will not produce a satisfactory finish unless the surface has been properly prepared. The basic principles are simple. They vary somewhat with different surfaces and, to some extent, with different paints; but the goal is the same—to provide a surface with which the paint can make a strong, permanent bond.

In General:

- The surface must be clean, smooth and free from loose particles such as dust or old paint. Use sandpaper, a wire brush or a scraper.
- Oil and grease should be removed by wiping with mineral spirits. If a detergent is used it should be followed by a thorough rinse with clean water.
- Chipped or blistered paint should be removed with sandpaper, a wire brush, steel wool or a scraper.
- Chalked or powdered paint should be removed with a stiff bristle brush, or by scrubbing with water mixed with household washing soda or TSP (trisodium phosphate, sold in hardware stores). If the old surface is only moderately chalked and the surface is relatively firm, an oil primer can be applied without the prior use of a stiff brush. The primer rebinds the loose articles and provides a solid base for the paint.
- Loose, cracked or shrunken putty or caulk should be removed by scraping.
- If new putty, glazing compound, caulking compounds and sealants are used, they should be applied to a clean surface and allowed to harden before paint is applied. If the caulk is a latex type, latex paint can be applied over it immediately without waiting for the caulk to harden.
- Damp surfaces must be allowed to dry before paint is applied, unless you are using a latex paint.

Wood Surfaces

- Scrape clean all areas where sap (resin) has surfaced on the wood, and sand smooth prior to application of "knot sealer." Small, dry knots should also be scraped and thoroughly cleaned, and then given a thin coat of knot sealer before applying wood primer.
- Fill cracks, joints, crevices and nail holes with glazing compound, putty or plastic wood and sand lightly until flush with the wood. Always sand in the direction of the grain—never across it.
- New wood surfaces to be stain-finished should first be sanded smooth. Open grain (porous) wood should be given a coat of paste filler before the stain is applied (paste fillers come in various matching wood colors). The surface should then be resanded. Read manufacturer's instructions carefully before applying paste fillers.

Masonry Surfaces

- Surfaces such as plaster, gypsum, cement and drywall should be dry and clean. If the surface is cracked, sand it smooth and then fill with spackling compound or some other recommended crack filler. After the repaired surface is dry, sand lightly until smooth—then wipe clean.
- Allow new plaster to dry for 30 days before painting.

- Roughen unpainted concrete and stucco with a wire brush to permit a good bond between the surface and the paint.
- Wash new concrete surfaces with detergent and water to remove any film left over from oil, or from the compound used for hardening the concrete during the "curing" process.
- Remove "efflorescence," the crystalline deposit which appears on the mortar between the bricks in a brick wall, by using undiluted vinegar or a 5% muriatic acid solution. After scrubbing with acid, rinse the surface thoroughly. CAUTION: When using muriatic acid, wear goggles and gloves for protection.

Metal Surfaces

- Clean new metal surfaces such as galvanized steel, aluminum or tin with a solvent such as mineral spirits to remove the oil and grease applied to the metal as a preservative by manufacturers.
- Remove rusted or corroded spots by wire-brushing or with coarse sandpaper. Chemical rust removers are also available from paint and hardware stores. Paint will not adhere well when applied over rusted or corroded surfaces.
- Allow galvanized steel, such as that used for roof gutters, to weather for about six months before painting. If earlier painting is necessary, wash the surface with mineral spirits or VM&P (Varnish Makers and Painters) naphtha, than apply a primer recommended specifically for galvanized surfaces.

CHECK THE WEATHER

You can easily ruin your paint job if you forget to consider the weather. Excessive humidity or extremely cold weather can cause you trouble. Good ventilation, regardless of the weather, is essential.

- Unless you are using latex paint, you should not paint on damp days. Moisture on the painting surface may prevent a good bond.
- If humidity is high, check the surface before painting. If you can feel a film of moisture on the surface, it would be better to wait for a better day. If you are painting inside and the area is air-conditioned, however, neither rain nor humidity will affect the job.
- Exterior painting is not recommended if the temperature is below 50 degrees or above 95 degrees Fahrenheit, since you may not be able to get a good bond. This is especially critical if you are using latex paint.
- If conditions are borderline, good ventilation will help paint to dry. Allow more drying time in damp or humid weather. The label on the can will tell you the normal drying time, but test each coat by touch before you add another. When paint is thoroughly dry, it is firm to the touch and is not sticky.

PAINTING INTERIOR SURFACES

Interior painting and exterior painting are similar in some ways but different in others. Because of the differences, the two types of painting will be treated separately. Some repetition will be unavoidable, but it will be kept to a minimum.

Interior Painting

Previously painted surfaces usually do not require primer coats except where the old paint is worn through or the surface has been damaged.

Wood Surfaces

- Unfinished wood to be finished with enamel or oil base paint should be primed with enamel undercoat to seal the wood and provide a better surface. If the unpainted wood is not primed, the enamel coat may be uneven.
- Unpainted wood to be finished with topcoat latex should first be undercoated. Water-thinned paint could raise the grain of the bare wood and leave a rough surface.
- If clear finishes are used: *Soft woods* such as pine, poplar and gum usually require a sealer to control the penetration of the finish coats. In using stain, a sealer is sometimes applied first in order to obtain a lighter, more uniform color; *Open grain hard woods* such as oak, walnut and mahogany require a paste wood filler, followed by a clear wood sealer; *Close grain hard woods* such as maple and birch do not require a filler. The first coat may be a thinned version of the finishing varnish, shellac or lacquer.

Masonry Surfaces

- Smooth, unpainted masonry surfaces such as plaster, plasterboard and various drywall surfaces can be primed with latex paint or latex primer-sealer. The color of the first coat should be similar to the finish coat.
- Coarse, rough or porous masonry surfaces, such as cement block, cinder block and concrete block cannot be filled and covered satisfactorily with regular paints. Block filler should be used as a first coat to obtain a smooth sealed surface over which almost any type of paint can be used.
- Unpainted brick, while porous, is not as rough as cinder block and similar surfaces and can be primed with latex primer-sealer or with an exterior-type latex paint.
- Enamel undercoat should be applied over the primer where the finish coat is to be a gloss or semi-gloss enamel.
- Follow carefully the manufacturer's instructions for painting masonry surfaces.

Metal Surfaces

- Unpainted surfaces should be primed for protection against corrosion and to provide a base for the finish paint. Interior paints do not usually adhere well to bare metal surfaces, and provide little corrosion resistance by themselves.
- Primer paints for bare metal surfaces must be selected according to the type of metal to be painted. Some primers are made especially for iron or steel; others for galvanized steel, aluminum or copper.
- An enamel undercoat should be used as a second primer if the metal surface is to be finished with enamel; that is, apply the primer first, then the

undercoat, and finally the enamel finish. Most enamel undercoats need a light sanding before the topcoat is applied.

Interior Paints and Finishes

Unless you are an experienced painter, shop for a salesman or a paint store owner before you shop for paint. Find one who is willing and able to help you match the paint to the job. Read labels and company leaflets carefully. They are usually well-written, accurate and helpful.

Paints for Light Wear Areas

- Latex interior paints are generally used for areas where there is little need for periodic washing and scrubbing; for example, living rooms, dining rooms, Bedrooms and closets.
- Interior flat latex paints are used for interior walls and ceilings since they cover well, are easy to apply, dry quickly, are almost odorless and can be quickly and easily removed from applicators.
- Latex paints may be applied directly over semi-gloss and gloss enamel if the surface is first roughened with sandpaper or liquid sandpaper. If the latter is used, follow carefully the instructions on the container label.
- Flat alkyd paints are often preferred for wood, wallboard and metal surfaces since they are more resistant to damage; also, they can be applied in thicker films to produce a more uniform appearance. They wash better than interior latex paints and are nearly odorless.

Paints for Heavy Wear Areas

- Enamels, including latex enamels, are usually preferred for kitchen, bathroom, laundry room and similar work areas because they withstand intensive cleaning and wear. They form especially hard films, ranging from flat to a full gloss finish.
- Fast-drying polyurethane enamels and clear varnishes provide excellent hard, flexible finishes for wood floors. Other enamels and clear finishes can also be used, but unless specifically recommended for floors they may be too soft and slow-drying, or too hard and brittle.
- Polyurethane and epoxy enamels are also excellent for concrete floors. For a smooth finish, rough concrete should be properly primed with an alkali resistant primer to fill the pores. When using these enamels, adequate ventilation is essential for protection from flammable vapors.

Clear Finishes for Wood

- Varnishes form durable and attractive finishes for interior wood surfaces such as wood paneling, trim, floors and unpainted furniture. They seal the wood, forming tough, transparent films that will withstand frequent scrubbing and hard use, and are available in flat, semi-gloss or satin, and gloss finishes.
- Most varnishes are easily scratched, and the marks are difficult to conceal without redoing the entire surface. A good paste wax applied over the finished varnish—especially on wood furniture—will provide some protection against scratches.
- Polyurethane and epoxy varnishes are notable for durability and high resistance to stains, abrasions, acids and alkalis, solvents, strong cleaners,

fuels, alcohol and chemicals. Adequate ventilation should be provided as protection from flammable vapors when these varnishes are being applied.

- Shellac and lacquer have uses similar to most varnishes, and these finishes are easy to repair or recoat. They apply easily, dry fast, and are also useful as a sealer and clear finish under varnish for wood surfaces. The first coat should be thinned as recommended on the container, then sanded very lightly and finished with one or more undiluted coats. Two coats will give a fair sheen, and three a high gloss.

Wax Finishes

- Liquid and paste waxes are used on interior surfaces. They provide a soft, lustrous finish to wood and are particularly effective on furniture and floors. Waxes containing solvents should not be used on asphalt tile; wax emulsions are recommended.
- Waxes should be applied to smooth surfaces with a soft cloth. Rub with the grain. Brushes should be used to apply liquid waxes to raw-textured wood.
- Wax finishes can be washed with a mild household detergent, followed by rinsing with a clean, damp cloth.
- A wax finish is not desirable if a different type of finish may be used later, for wax is difficult to remove.

COLOR DO'S AND DON'TS

DO use light colors in a small room to make it seem larger.

DO aim for a continuing color flow through your home—from room to room—using harmonious colors in adjoining areas.

DO paint the ceiling of a room in a deeper color than walls, if you want it to appear lower; paint it in a lighter shade for the opposite effect.

DO study color swatches in both daylight and nightlight. Colors often change under artificial lighting.

DON'T paint woodwork and trim of a small room in a color which is different from the background color, or the room will appear cluttered and smaller.

DON'T paint radiators, pipes and similar projections in a color which contrasts with walls or they will be emphasized.

DON'T choose neutral or negative colors just because they are safe, or the result will be dull and uninteresting.

DON'T use glossy paints on walls or ceilings of living areas since the shiny surface creates glare.

WHAT TO USE AND WHERE
(Interior Surfaces)

	Latex Gloss & Semi-gloss	Latex (Wall) Flat	Metal Primer	Sealer or Undercoater	Aluminum Paint	Floor Paint or Enamel	Floor Varnish	Wood Sealer	Stain	Wax (Emulsion)	Wax (Liquid or Paste)	Shellac - Lacquer	Interior Varnish	Gloss Enamel	Semi-gloss Enamel	Flat Enamel
MASONRY																
Asphalt Tile										X						
Concrete Floors		X				X			X	X.	X.					
Kitchen & Bathroom Walls	X.			X										X.	X.	
Linoleum										X						
New Masonry	X.	X		X											X.	X.
Old Masonry	X.	X		X	X										X	X
Plaster Walls & Ceiling	X.	X		X											X.	X.
Vinyl & Rubber Tile Floors										X	X					
Wall Board	X.	X		X											X.	X.
METAL																
Aluminum Windows	X.	X.	X		X										X.	X.
Heating Ducts	X.	X.	X		X										X.	X.
Radiators & Heating Pipes	X.	X.	X		X										X.	X.
Steel Cabinets	X.		X												X.	X.
Steel Windows	X.	X.	X		X										X.	X.
WOOD																
Floors						X.	X.	X	X.	X.	X	X				
Paneling	X.	X.						X	X		X	X	X		X.	X.
Stair Risers	X.							X	X			X	X		X.	X.
Stair Treads						X	X	X	X			X				
Trim	X.	X.		X					X		X	X	X		X.	X.
Window Sills													X			

[X.] Black dot indicates that a primer or sealer may be necessary before the finishing coat (unless surface has been previously finished).

INTERIOR PAINT

Properties and Uses

Types	Properties	Typical Uses
Latex primer-sealer (water thinned)	Simple to apply. Dries quickly and can be recoated in about two hours. Not flammable; almost odorless. One coat usually sufficient. Thinning unnecessary unless recommended by manufacturer.	Unpainted interior walls and ceilings of wallboard, plaster, masonry and all types of drywall.
Enamel undercoater (alkyd base - low odor type)	Hard, tight films. Provides good base for enamel. Easy brushing, smooth leveling. Dries in about 12 hours.	Undercoater for interior enamels.
Latex wall paint (water thinned)	The most popular of the interior paints. Durable, excellent coverage, good washability, quick-drying, and easy to touch-up. Safe to use and store; nontoxic, practically no odor.	Primer-sealer and also finish coat for interior walls and ceilings of wallboard, wallpaper, plaster, and other porous, absorptive materials. Use on primed wood but not on bare wood.
Flat alkyd enamel	Made with alkyd resins. Has flat finish practically free of sheen. Used same as latex wall paint but has slightly better washability and abrasion resistance. Dries in about four hours. Practically no odor.	Primer-sealer and also finish coat on interior walls and ceilings of plaster, wallboard, masonry, and similar surfaces.
Semi-gloss and full-gloss enamel (alkyd base)	Made with alkyd resins. Has good gloss retention, grease and oil resistance, and better washability and resistance to abrasion than flat alkyd enamel.	On primed plaster and wallboard, and on suitably prepared wood trim and metal. Very useful for kitchens and bathrooms, and for decorative use on properly primed woodwork.
Semi-gloss and full-gloss latex enamel (water thinned)	Has most properties of alkyd enamels plus usual advantages of latex paints: easy application and cleanup, rapid drying, low odor, and nonflammable. Good leveling but lapping does not compare favorably with alkyd enamels.	Walls and ceilings of wallboard, wallpaper, wood, and plaster. Very useful for kitchens, bathrooms, and for decorative use on properly primed woodwork.
Epoxy enamel	Hard film, wide gloss range, low odor, ideal where vigorous and frequent cleaning is done. Has excellent adhesion and resistance to abrasion, water, solvents, greases, and dirt. Packaged in two containers -- enamel in one and curing agent in the other. Contents of both containers mixed together prior to use. Cost comparatively high but durability is excellent.	Highly effective in heavy wear areas such as hallways, kitchens, bathrooms, laundries and concrete floors.
Dripless enamel (special alkyd base)	Does not drip from brush or roller. Made with special alkyd resins to form a soft gel which liquefies with agitation but gels again on standing. Soft, buttery, easy brushing, low odor, self-sealing. Has excellent color retention; solvent and water resistant.	Decorative enamel for properly primed walls and ceilings of plaster, wallboard, and similar surfaces; also, for wood trim and primed metal.

Types	Properties	Typical Uses
Interior floor and deck enamel	Alkyd and latex used successfully but polyurethane types provide harder, more flexible and more abrasion-resistant surfaces. Polyurethane enamels are packaged in one and two-container forms. The latter has paint in one can and curing agent in the other. Contents of both are mixed together prior to use. When applying polyurethane enamel, follow manufacturer's instructions explicitly and also keep room well ventilated.	General application to properly primed floors and covered decks.
Clear varnish finishes for wood	Provide durable and attractive finish; seal wood better than lacquer; and form tough, transparent coat that will withstand frequent scrubbings and hard use. Tend to darken the wood surface and give impression of visual depth. Readily show scratch marks which are difficult to conceal without redoing entire surface. Some varnishes turn yellow with age. Extra coat recommended on new work. Can be flat, satin, semi-gloss, or glossy finish.	For all interior smooth wood. Recommended for washrooms, kitchens, or other areas exposed to dirt, grease, and moisture and subject to frequent scrubbing. A rubbed-in coat of paste wax will provide some protection against scratches.
Shellac	Available in clear and "orange" finishes. Fast drying. Thinned first coat provides excellent seal for new wood. Can be over-coated in about 30 minutes. Should be lightly sanded between coats. Paste wax, as final coat, provides lustre and some protection against scratches.	For wood walls, trim, furniture, or any wood surface requiring only occasional dusting. Unsuitable for kitchens, washrooms, or other areas exposed to dirt, grease, and moisture.
Lacquer	Fast drying; can be overcoated in about 30 minutes. Provides gloss or sheen when two or more coats are applied. Paste wax, as final coat, provides attractive lustre and some protection against scratches. Available in clear and a variety of color finishes.	For wood walls, trim, furniture, or any wood surface requiring only occasional dusting. Unsuitable for kitchens, washrooms, or other areas exposed to dirt, grease, and moisture.
Stains	Available in natural finish and in a variety of colors which provide attractive, natural appearance. Several coats are required for bare wood, with light sanding between coats. Final coat of paste wax provides lustre and some protection against scratches, particularly on furniture. "Thick" stains can be thinned with turpentine or mineral spirits.	For interior wood surfaces such as walls, trim, and furniture.
Aluminum paints	Resistant to water. Can be brushed or sprayed on new metal and wood surfaces. When brush is used, apply in one direction only for best results.	As a sealer for wood surfaces (especially knots) and as a primer for metal surfaces. Can be used as a finish coat if color is not objectionable. Particularly useful for aluminum and steel windows, heating ducts, radiators and heating pipes.

PAINTING EXTERIOR SURFACES

The durability of an exterior paint job depends greatly on surface preparation, the quality of paint selected, the skill of application, the proper spacing of repaintings, the protection of surfaces from the sun and rain, and climactic and local weather conditions.

As previously indicated, conditions must be right for exterior painting. The temperature should not be much below 50 degrees or above 95 degrees Fahrenheit, and surfaces must be free from moisture. Latex paints, however, can be used even if the surface is not bone dry. The best time for exterior painting is after morning dew has evaporated.

Before you start on the job, make a thorough inspection tour and check the surface condition of window and door frames and surrounding areas, bases of columns of porches and entranceways, steps, siding, downspouts, under-eave areas, and anywhere that moisture is likely to collect.

EXTERIOR PRIMING

Wood Surfaces

- The tendency of wood to expand and contract during changes in temperature and humidity makes it imperative that a good wood primer be applied to provide the necessary anchorage for the finish paint.
- Surfaces such as wood siding, porches, trim, shutters, sash doors and window sills should be primed with an exterior primer intended for wood. Application should be by brush to thoroughly dry surfaces.
- Painted wood usually does not need priming unless the old paint has cracked, blistered or peeled. Defective paint must be removed by scraping or wire brushing—preferably down to bare wood—and then primed.
- Scratches, dents, recesses and raw edges should be smoothed and then touched up with a suitable exterior primer.

Masonry Surfaces

- New masonry surfaces should be primed with an exterior latex paint, preferably one made specifically for masonry.
- Common brick is sometimes sealed with a penetrating type of clear exterior varnish to control efflorescence and spalling (flaking or chipping of the brick). This varnish withstands weather, yet allows the natural appearance of the surface to show through.
- Coarse, rough and porous surfaces should be covered with a fill coat (block filler), applied by brush to thoroughly penetrate and fill the pores.
- Old painted surfaces which have become a little chalky should be painted with an exterior oil primer to rebind the chalk. If there is much chalk, it should be removed with a stiff bristle brush or by washing with household washing soda or TSP (trisodium phosphate) mixed with water.

Metal Surfaces

- Copper should be cleaned with a phosphoric acid cleaner, buffed and polished until bright, and then coated before it discolors. Copper gutters and downspouts do not, however, require painting. The protective oxide which forms on the copper surface darkens it or turns it green, but does not shorten the life of the metal. Copper is often painted to prevent staining of adjacent painted surfaces.

- Zinc chromate type primers are effective on copper, aluminum and steel surfaces, but other types are also available for use on metal.
- Galvanized steel surfaces, such as gutters and downspouts, should be primed with recommended special primers since conventional primers usually do not adhere well to this type of metal. A zinc-dust zinc-oxide type primer works well on galvanized steel. Exterior latex paints are sometimes used directly over galvanized surfaces, but not oil paints.
- Unpainted iron and steel surfaces rust when exposed to the weather. Rust, dirt, oils and loose old loose paint should be removed from these surfaces by wire brushing or power tool cleaning. The surface should then be treated with an anti-corrosive primer.

EXTERIOR FINISHING

- All exterior surfaces, properly primed or previously painted, can be finished with either exterior oil paint or exterior latex paint.
- Latex paints are easy to apply, have good color retention, and can be used on slightly damp surfaces.
- Oil or alkyd base paints have excellent penetrating properties. They provide good adhesion, durability and resistance to abrasion and blistering on wood and other porous surfaces.
- Mildew, fungus and mold growths on exterior surfaces are a problem in areas where high temperature and humidity are prevalent. Use paint which contains agents to resist bacterial and mold growth. The manufacturer's label will state whether the paint contains such inhibitors.
- Exterior oil and latex paints can be applied by brush or by spraying; however, brush application generally provides a more intimate bond between surface and the paint film.
- Colored exterior house paints must resist chalking so that colors will not fade and the erosion of the paint film will be minimized. The manufacturer's label will indicate whether the paint is a non-chalking type. Some white exterior house paints are expected to chalk slightly as a means of self-cleaning.

WHAT TO USE AND WHERE
(Exterior Surfaces)

	Water Repellent Preservative	House Paint (Latex)	Metal Primer	Primer or Undercoater	Porch and Deck Paint	Trim Paint	Roof Coating	Wood Stain	Aluminum Paint	Exterior Clear Finish	Cement Powder Paint	House Paint (Oil or Oil-Alkyd)
MASONRY												
Asbestos Cement		X		X								X•
Brick	X	X		X					X		X	X•
Cement & Cinder Block		X		X					X		X	X•
Concrete/Masonry Porches And Floors		X			X							
Coal Tar Felt Roof							X					
Stucco		X		X					X		X	X•
METAL												
Aluminum Windows		X•	X			X•			X			X•
Steel Windows		X•	X			X•			X•			X•
Metal Roof		X•	X									X•
Metal Siding		X•	X			X•			X•			X•
Copper Surfaces										X		
Galvanized Surfaces		X•	X			X•			X•			X•
Iron Surfaces		X•	X			X•			X•			X•
WOOD												
Clapboard		X•		X					X			X•
Natural Wood Siding & Trim								X		X		
Shutters & Other Trim		X•		X		X•						X•
Wood Frame Windows		X•		X		X•			X			X•
Wood Porch Floor					X							
Wood Shingle Roof	X							X				

X• Black dot indicates that a primer, sealer, or fill coat may be necessary before the finishing coat (unless surface has been previously finished).

EXTERIOR PAINT

Properties and Uses

Types	Properties	Typical Uses
Oil base primers	Good adhesion and sealing; resistant to cracking and flaking when applied to unprimed wood; good brushing and leveling; controlled penetration; and low sheen. Unsuitable as a top coat and should be covered with finish paint within a week or two after application.	As primer on unpainted woodwork or surfaces previously coated with house paint.
Anti-rust primers	Prevent corrosion on iron and steel surfaces. Slow-drying type provides protection through good penetration into cracks and crevices. Fast-drying types are used only on smooth, clean surfaces, and those which are water resistant are effective where surfaces are subject to severe humidity conditions or fresh water immersion.	Priming of steel and other ferrous metal surfaces when good resistance to corrosion is required.
Galvanizing primers	High percentage of zinc dust provides good anti-rust protection and adhesion. Galvanizing/zinc dust primers give excellent coverage, one coat usually being sufficient on new surfaces. Two coats are ample for surfaces exposed to high humidity.	Priming of new or old galvanized metal and steel surfaces. Satisfactory as finish coat if color (metallic gray) is not objectionable.
House paints (oil or oil alkyd base)	Made with drying oils or drying oil combined with alkyd resin. Excellent brushing and penetrating properties. Provides good adhesion, elasticity, durability, and resistance to blistering on wood and other porous surfaces. Often modified with alkyd resins to speed drying time. Apply with brush to obtain strong bond, especially on old painted surfaces.	General exterior use on properly primed or previously painted wood or metal surfaces.
House paints (latex type)	Exterior latex paints have durability comparable to oil base paints. Resistant to weathering and yellowing, and so quick-drying that they can be recoated in one hour. Can be applied in damp weather over a damp surface. Easy to apply and brush or roller can be cleaned quickly with water. Free from fire hazard. White latex paints usually offer better color retention than oil or oil-alkyd exterior paints.	Covers properly primed or previously painted concrete, stucco, and other masonry and wood surfaces.
Trim paint	Usually made with oil modified alkyds. Slow drying (over night). Made in high sheen, bright colors; have good retention of gloss and color. More expensive silicone-alkyd enamels are also available for trim painting. They are substantially more durable than conventional oil-alkyd enamels.	Applied over primed wood and metal surfaces such as aluminum and steel windows, metal siding, shutters and other trim, and wood frame windows.

Types	Properties	Typical Uses
Porch and deck paints (for concrete and other masonry surfaces)	Many made with natural rubber base (chlorinated rubber resin). Good flow and leveling; good resistance to rain, moisture, and detergents. Coverage about 300 square feet per gallon. New concrete should age at least two months before painting. Three coats recommended, including thinned first coat. Good ventilation required if used indoors.	For both interior and exterior concrete and masonry porches and decks. Effective for swimming pools, shower rooms, and laundries. Manufacturer's instructions must be followed carefully.
Porch and deck paints (for wood surfaces)	A variety of alkyd base and other types available. Tough, flexible, and abrasion resistant. Good drying properties. Thin coats promote thorough drying. Allow ample drying time between coats.	Interior and exterior decks and porches.
Aluminum paints	Resistant to water and weather and provide excellent durability in marine environments. Can be applied on new metal or wood surfaces — in one direction for best results. Creosote-treated wood must age for about six months prior to application of aluminum paint.	Particularly useful in marine environments; as a sealer for wood knots; and as a combination sealer and finish coat for wood surfaces treated with creosote or other preservatives. Can also be used as a finish coat for metal and wood if aluminum color is not objectionable.
Cement powder paint	Made from white Portland cement, pigments, and (usually) small amounts of water repellent. They are mixed with water just before application. Painted surfaces should be kept damp by sprinkling with water until paint film is well cured. This paint does not provide a good base for other types of finishes. Apply with a fiber brush.	Useful, low cost finish for rough masonry surfaces, both interior and exterior, including brick, cement and cinder block, and stucco.
Wood stains	Semi-transparent type available for exterior wood but not as durable as house paints. Improves appearance of wood by highlighting the grain and texture of the surface. Available in many colors, the most popular being cedar, light redwood, and dark redwood.	For smooth and rough wood surfaces. Can be used for staining house siding and wood fencing, but should not be used on surfaces that may soon be painted or on previously painted surfaces. NOT recommended for frames, windows, and doors which need a high degree of protection against the weather.

Types	Properties	Typical Uses
Clear finishes (for wood)	Not as durable as pigmented paint. Alkyd varnishes have good color and color retention but may crack and peel. Some synthetics such as polyurethane varnishes have good durability but may darken on exposure. Spar varnish (marine varnish) is quite durable but will also darken and yellow. Use thin penetrating coats on the bare wood, followed by the unreduced varnish.	For clear finish on wood surfaces where natural appearance is desired.
Roof coatings	Bituminous roof coatings are made of asphalt (chosen for good weather resistance), dissolved in a suitable solvent. Asbestos and other fillers are added to prevent sagging on sloping roofs and to permit application of relatively thick coatings. Basically made in gray and black; however, addition of aluminum powders provides for other colors. Asphalt emulsion roof coatings can be applied over damp surfaces. Special application techniques are usually required and manufacturer's instructions must be followed carefully.	Used primarily for coal tar felt roofs.
Water repellent preservative (silicone type)	Silicone water repellents are transparent liquids that help repel water without changing the surface appearance. Must be applied strictly in accordance with instructions to ensure adequacy of film and water repellency. Should not be topcoated with paint until surface has weathered for at least two years.	For wood shingle roofs, brick walls, and other surfaces where some degree of water repellancy is desired.

SUGGESTED COLOR SCHEMES

If your house has shutters, paint the trim the same color as the body of the house—or white. If not, use these suggested colors for trim.

If the roof of your house is	You can paint the body	. . . and the trim or shutters and doors															
		Pink	Bright red	Red-orange	Tile red	Cream	Bright yellow	Light green	Dark green	Gray-green	Blue-green	Light blue	Dark blue	Blue-gray	Violet	Brown	While
GRAY	White	x	x	x	x	x	x	x	x	x	x	x	x	x	x		
	Gray	x	x	x	x		x	x	x	x	x	x	x	x	x		x
	Cream-yellow		x		x		x		x	x							x
	Pale green				x		x		x	x							x
	Dark green	x				x	x	x									x
	Putty			x	x				x	x			x	x		x	
	Dull red	x				x		x						x			x
GREEN	White	x	x	x	x	x	x	x	x	x	x	x	x	x	x	x	
	Gray			x		x	x	x									x
	Cream-yellow		x		x			x	x	x						x	x
	Pale green			x	x		x		x								x
	Dark green	x		x		x	x	x									x
	Beige				x				x	x	x		x	x			
	Brown	x				x	x	x									x
	Dull red					x		x		x							x
RED	White		x		x				x		x			x			
	Light gray		x		x				x								x
	Cream-yellow		x		x						x		x	x			
	Pale green		x		x												x
	Dull red					x		x		x	x						x
BROWN	White			x	x		x	x	x	x	x		x	x	x	x	
	Buff				x				x	x	x					x	
	Pink-beige				x				x	x						x	x
	Cream-yellow				x				x	x	x					x	
	Pale green								x	x						x	
	Brown			x		x	x										x
BLUE	White			x	x		x					x	x				
	Gray			x		x						x	x				x
	Cream-yellow			x	x								x	x			
	Blue			x		x	x					x					x

CAUSES OF PAINT FAILURES

		Causes	Treatment
	~~**Blistering**~~	~~Inside and outside moisture from~~ poor construction; e.g., unpainted openings, poor ventilation, inadequate insulation. Cooking, bathing, dishwashing, laundry, etc. (Daily evaporation in average home: 25 quarts.)	Caulk and repair gutters, faulty siding, and roof flashings. Repair leaky roof. Provide attic louvres, exhaust fans, dehumidifiers, all possible ventilation. Scrape blisters; repaint with blister resistant paint.
	**Excessive Chalking**	Paint applied in rain, fog, or mist. Paint applied too thin. Low quality paint.	Remove chalk with a stiff bristle brush. Use nonchalking paint, or one with controlled rate of chalking if desirable to maintain clean surfaces.
	**Alligatoring**	Low quality paint. Insufficient drying time between coats. A hard coating applied over a soft oil-base type paint.	Scrape thoroughly and sand smooth. Apply good quality paint according to manufacturer's instructions.
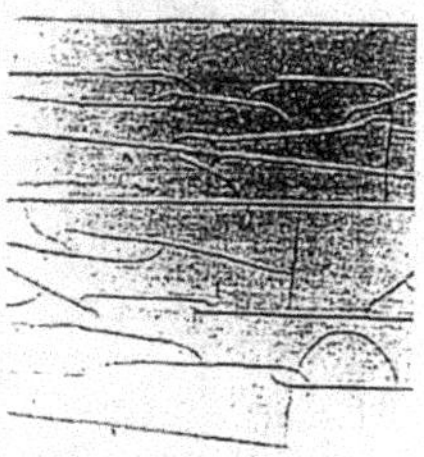	**Checking**	Oil paint applied over damp surface. Low quality paint. Improperly mixed paint. Unevenly applied paint. Excessive paint.	Scrape paint to bare surface. Apply good quality paint over a clean, primed surface.
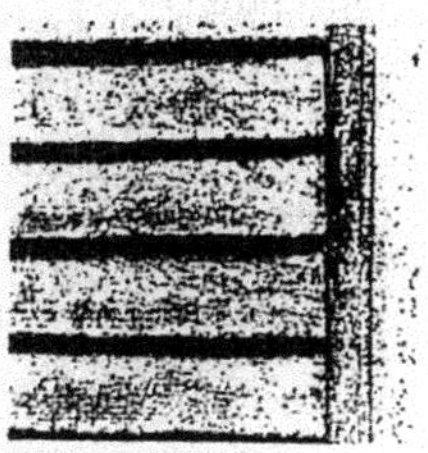	**Mildew**	Mildew producing fungi. Insufficient sunlight on surfaces in damp areas.	Apply solution of trisodium phosphate (available in paint stores) mixed with household ammonia and water, followed by clean water rinse - OR wash down with a solution of 1 pint of household laundry bleach to 1 gallon of water, followed by a thorough clean water rinse. Use mildew resistant paint containing zinc oxide and a mildewcide.
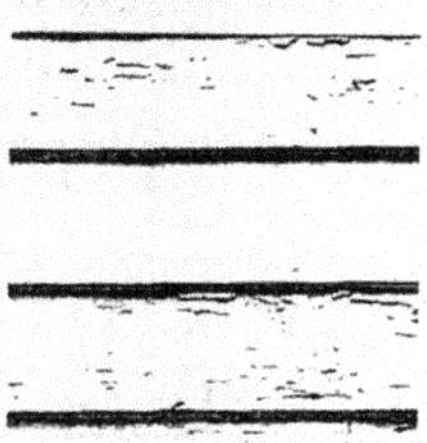	**Peeling**	Paint applied over greasy or oily surface. Inadequately prepared and primed surfaces. Application of oil-base paint over damp surfaces.	Scrape and sand peeled surfaces. Apply good quality paint, with brush.

FIGURING YOUR PAINT ORDER

After you have a clear idea of how to go about preparing the surface, priming the surface and applying the paint, the next step is to decide on the kind of paint or finish to use and how much to buy.

The Information in this guide will help you decide on the general type of finish for the purpose you have in mind, but it will be good to discuss the project with a paint dealer in your vicinity whom you know, or who has a good reputation in the community. He can help you avoid the common mistakes painters make when they are learning. He can also help you select a specific paint for the job and tell you how much you need.

For a rough estimate, you can figure the amount of paint you need for flat surfaces by simply multiplying the height (or length) times the width and dividing the result into the coverage estimate on the label. If, for example, you wish to paint a room which has 416 square feet of wall area, a gallon of paint advertised as covering 500 square feet will be adequate for one coat. Second coats generally require less paint.

But remember that these are *average* estimates of coverage. Some surfaces are more absorbent than others, and your paint dealer can give you the benefit of his experience in tailoring an estimate to the particular surfaces you are going to be covering. Be ready to provide him with exact measurements of the areas to be covered.

Your dealer will also be able to advise you about the number of coats that will be required for different surfaces and different types of paint. <u>Don't buy until you and the dealer have reached agreement on the approximate amount of paint, thinner, primer and other supplies you need.</u> If a particular dealer is lacking in patience in reaching this agreement, find another dealer.

PAINTING CHECKLIST

Before starting to paint, make sure you have all the tools needed to do the job right. The following are some of the items usually required:

Paint brushes
Stiff bristle brush
Caulking gun
Drop cloths
Hammer
Masking tape
Paint
Paint scraper
Cheesecloth
Patching plaster
Putty knife
Rollers
Sandpaper or production paper
Steel wool
Wire comb

Dust brush
Wire brush
Cans*
Emery cloth**
Ladder
Mixing paddle
Paint bucket
Paint strainer/wire mesh
Paint tray***
Putty or glazing compound
Rags
Roller extension poles
Spackling compound
Turpentine or other solvents

* For cleaning brushes with solvents
**For cleaning and polishing metal surfaces
***For painting with rollers

BASIC FUNDAMENTALS OF PAINT AND INTERIOR PAINTING

CONTENTS

BASIC FUNDAMENTALS OF PAINT AND INTERIOR PAINTING

For an attractive, long-lasting paint job, you need to:

1. Use good-quality paint,
2. Properly prepare the surface for painting, and
3. Apply the paint correctly.

Preparation of the surface—cleaning and patching—may take the most time in painting, but it is the most important part of the job. Even the best paint will not adhere well to an excessively dirty or greasy surface or hide large cracks or other mars.

PAINT SELECTION

Kind

Many different kinds and formulations of paints and other finishes are available for interior use, and new ones frequently appear on the market.

Use the chart on the next page as a general guide in making your selection. For a more specific selection, consult your paint dealer. Reputable paint dealers keep abreast of the newest developments in the paint industry and stock the newest formulations.

Dripless paint is an example of a fairly recent development. It has a jelled consistency in the can, but it loses that form when picked up on a brush or roller and spreads evenly and smoothly. It is particularly convenient when painting a ceiling.

The usual interior paint job consists of painting wallboard or plaster walls and ceilings, woodwork, and wood windows and doors. For these surfaces, you need to choose first between solvent-thinned paint (commonly called oil-based paint) and water-thinned paint (commonly called latex paint, but not necessarily latex), and then between a gloss, semigloss, or flat finish.

Enamels, which are made with a varnish or resin, base instead of the usual linseed oil vehicle, are included under the broad oil-paint grouping.

Oil-based paints are very durable, are highly resistant to staining and damage, can withstand frequent scrubbings, and give good one-coat coverage. Many latex paints are advertised as having similar properties.

The main advantages of latex paint are easier application, faster drying, and simpler tool cleanup. The brushes, rollers, and other equipment can be easily cleaned with water.

Both oil-based paint and latex paint are now available in gloss, semigloss, and flat finishes. Glossy finishes look shiny and clean easily. Flat finishes show dirt more readily but absorb light and thus reduce glare. Semigloss finishes have properties of both glossy and flat finishes.

GUIDE FOR SELECTING PAINT

	Aluminum paint	Casein	Cement base paint	Emulsion paint (including latex)	Enamel	Flat paint	Floor paint or enamel	Floor varnish	Interior varnish	Metal primer	Rubber base paint (not latex)	Sealer or undercoater	Semigloss paint	Shellac	Stain	Wax (emulsion)	Wax (liquid or paste)	Wood sealer
Floors:																		
Asphalt tile																	x•	
Concrete																x•	x•	
Linoleum							x							x		x	x	
Vinyl and rubber							x	x								x	x	
Wood							x•	x•									x	
Masonry:																		
Old	x	x	x	x	x•	x•					x	x	x•					
New			x	x	x•	x•					x	x	x•					
Metal:																		
Heating ducts	x				x•	x•				x	x		x•					
Radiators	x				x•	x•				x	x		x•					
Stairs:																		
Treads							x	x						x	x			
Risers					x•	x•			x		x		x•	x	x			
Walls & Ceilings																		
Kitchen & bathroom				x	x•						x	x	x•					
Plaster		x		x		x•					x	x	x•					
Wallboard		x		x		x•					x	x	x•					
Wood paneling				x•		x•			x									
Wood trim				x•	x•	x•			x		x	x	x•	x	x		x	x
Windows:																		
Aluminum	x				x•	x•				x	x		x•					
Steel	x				x•	x•				x	x		x•					
Wood sill					x•				x			x			x			

Black dot (x•) indicates that a primer or sealer may be necessary before the finishing coat (unless the surface has been previously finished).

Because enamel is durable and easy to clean, semigloss or full-gloss enamel is recommended for woodwork and for the walls of kitchens, bathrooms, and laundry rooms. For the walls of nurseries and other playrooms, either oil-based or latex semigloss enamel paint is suggested. Flat paint is generally used for the walls of living rooms, dining rooms, and other non-work or non-play rooms.

Color

Paints are available in a wide range of colors and shades. Dealers usually carry color charts showing the different possibilities. Some of the colors are ready-mixed; others the dealer has to mix by adding or combining different colors.

Color selection is mostly a matter of personal preference. Here are some points to keep in mind in selecting your colors:

- Light colors make a small room seem larger. Conversely, dark colors make an overly large room appear smaller.

- Bright walls in a large room detract from otherwise decorative furnishings.

- Ceilings appear lower when darker than the walls and higher when lighter than the walls.

- Paint generally dries to a slightly different color or shade. For a fast preview of the final color, brush a sample swatch of the paint on a piece of clean, white blotting paper. The blotting paper will immediately absorb the wet gloss, and the color on the paper will be about the color of the paint when it dries on the wall.

- Colors often change under artificial lighting. Look at color swatches both in daylight and under artificial light.

Quantity

For large jobs, paint is usually bought by the gallon. The label usually indicates the number of square feet a gallon will cover when applied as directed. To determine the number of gallons you need:

1. Find the area of the walls in square feet by multiplying the distance around the room by the height of the walls. (This figure will include door and window space.)

2. From this figure, subtract *one-half* of the total area, in square feet, taken up by doors and windows. To find this area, multiply the height of each unit by its width, then add the results.

3. Divide the figure obtained in Step 2 by the number of square feet a gallon will cover. Then, multiply by the number of coats to be applied. The result is the number of gallons needed.

Ceilings are frequently painted a different color or shade (usually white) than the walls and need to be figured separately. To find the square feet area of the ceiling, multiply the length by the width.

Keep in mind that unpainted plaster and wallboard soak up more paint than previously painted walls and, therefore, require more paint or primer.

Some paints are guaranteed to give one-coat coverage over all or most colors if applied as directed at a rate not exceeding the number of square feet specified on the label of the paint container.

SURFACE PREPARATION

In general, walls, cellings, woodwork, and other surfaces to be painted should be clean, dry, and smooth. But read the label on the paint can before you start painting; it may contain additional or special instructions for preparing the surface.

Plaster and Wallboard

New Surfaces

New plaster walls should not be painted with oil-based paint until they have thoroughly cured—usually after about two months. Then, a primer coat should be applied first.

If necessary to paint uncured plaster, apply *one coat only* of a latex paint or primer. Latex, or water-base, paint will not be affected by the alkali in new plaster and will allow water to escape while the plaster dries. Subsequent coats of paint—either oil-based or latex—can be added when the plaster is dry.

Unpainted plaster readily picks up and absorbs dirt and it difficult to clean. The one coat of latex paint or primer will protect it.

For new drywall, a latex primer or paint is recommended for the first coat. Solvent-thinned paints tend to cause a rough surface. After the first coat of latex paint, subsequent coats can be of either type.

Clean or dust new surfaces before you apply the first coat of primer or paint.

Old Surfaces

The first step is to inspect the surface for cracks and mars. Fill small hairline cracks with spackling compound and larger cracks with special patching plaster. Follow the directions on the container label when using the patching material. When the patch is completely dry, sand it smooth and flush with the surrounding surface.

Nailheads tend to *pop out* in wallboard walls and ceilings. Countersink the projecting heads slightly and fill the hole with spackling compound. Sand the patch smooth when it is dry. It is desirable to prime newly spackled spots, particularly if you are applying only one coat.

Next, clean the surface of dirt and grease. A dry rag or mop will remove dust and some dirt. You may have to wash the surface with a household cleaner to remove stubborn dirt or grease.

Kitchen walls and ceilings are usually covered with a film of grease from cooking (which may extend to the walls and ceilings just outside the entrances to the kitchen), and bathroom walls and ceilings may have steamed-on dirt. The grease or dirt must be removed—the new paint will not adhere to it. To remove the grease or dirt, wash the surface with a strong household cleanser, turpentine, or mineral spirits.

The finish on kitchen and bathroom walls and ceiling is usually a gloss or semigloss. It must be *cut* so that the new paint can get a firm hold. Washing the surface with the household cleanser or turpentine will dull the gloss, but, for best results, rub the surface with fine sandpaper or steel wool. After using sandpaper or steel wool, wipe the surface to remove the dust.

Woodwork

Woodwork (windows, doors, and baseboards) usually have a glossy finish. First, wash the surface to remove dirt and grease, and then sand it lightly to *cut* the finish so that the new paint can get a good hold. After sanding, wipe the surface so that the new paint can get a good hold. After sanding, wipe the surface to remove the dust.

You can buy liquid preparations that will soften hard, glossy finishes to provide good adhesion for the new paint.

If there are any bare spots in the wood, touch up with an undercoater or with pigmented shellac before you paint.

APPLICATION

Read the label on the paint can before you start painting. It will contain general application instructions and may contain special instructions.

Equipment

Interior painting is usually done with brushes or with brushes and rollers. Indoor spray painting is not generally done by the homeowner, except for small jobs using pressurized cans of paint.

For speed and convenience, use a roller on the walls, ceilings, and other large surfaces, and then use a brush at corners, along edges, and in other places that you cannot reach with a roller. Woodwork is usually painted with a brush.

Special-shaped rollers and other applicators are available for painting woodwork, corners, edges, and other close places. Some may work fine; others, not so well. You may find that a small brush is still best for such work.

Different kinds of brushes and rollers are recommended for use with different kinds of paint. For example, short-nap rollers are best for applying gloss enamel on smooth surfaces. Check with your paint dealer on what kind of brush or roller to buy.

Other equipment needed for indoor painting includes a stepladder, drop cloths, and wiping rags.

SAFETY TIPS

For a safer paint job:

- Never paint in a completely closed room, and use caution when painting in a room where there is an open flame or fire. Some paints give off fumes that are flammable or dangerous to breathe or both.

  Avoid prolonged exposure to paint fumes for a day or two after painting. Such fumes can be especially harmful to canaries or other pet birds.

- Use a sturdy stepladder or other support when painting high places. Be sure that the ladder is positioned firmly, with the legs fully opened and locked in position.

- Face the ladder when climbing up or down it, holding on with at least one hand. Lean toward the ladder when painting.

- Do not overreach when painting. Move the ladder frequently rather than risk a fall. To avoid spilling the paint, take the few seconds required to remove the paint can from the ladder before you move it.

- When you finish painting, dispose of the used rags by putting them in a covered metal can. If left lying around, the oily rags could catch fire by spontaneous combustion.

- Store paint in a safe, but well-ventilated, area where children and pets cannot get to it. A locked cabinet is ideal if well-ventilated. Unless needed for retouching, small quantities of paint may not be worth saving.

PAINTING TIPS

For an easier and better paint job:

- Do the painting when the room temperature is comfortable for work—between 60° and 70°F, and provide good cross-ventilation both to shorten the drying time and to remove fumes and odors.

  Note: Check the label on the paint can for any special application and drying instructions.

- Preferably, remove all furnishings from the room. Otherwise, cover the furniture, fixtures, and floor with drop cloths or newspapers. No matter how careful you may be, you will spill, drip, or splatter some paint.

- Remove all light-switch and wall-plug plates. Paint the plates before you replace them after painting the room.

- Dip your brush into the paint no more than one-third the length of the bristles. This will minimize splattering and dripping.

- When using latex paint, wash your brush or roller occasionally with water. A buildup of the quick-drying paint in the nap of the roller or at the base of the bristles of the brush could cause excessive dripping.

- Wipe up spilled, splattered, or dripped paint as you go along. Paint is easier to cleanup when wet.

- Do not let the paint dry out in the can or in brushes or rollers between jobs or curing long interruptions in a job. After each job, replace the can lid, making sure that it is on tightly, and clean brushes or rollers. During long interruptions in a job, also replace the can lid, and either clean brushes or rollers or suspend them in water.

PROCEDURE

Paint the ceiling first. Don't try to paint too wide a strip at a time. The next strip should be started and lapped into the previous one before the previous one dries.

If you are putting two coats on the ceiling, apply the second coat and *cut in* at the junction with the walls, before you paint the walls.

Start painting a wall at the upper left-hand corner and work down toward the floor (left-handed persons may find it more convenient to start at the upper right-hand corner).

Paint the woodwork (windows, doors, and baseboards) last—preferably after the walls are completely dry.

Flush doors can be painted with a roller. On paneled doors, some parts can be painted with a roller; other sections will require a brush. You may prefer your doors and other trim in natural color. See the section entitled, Natural Finishes For Trim.

Paint the parts of a window in the order shown in the drawing below. Windows are easier to paint and to clean afterward if the glass is masked. Both masking tape and liquid masking are available at hardware and paint stores.

Paint windows in this order:

1. Mullions
2. Horizontal of sash
3. Verticals of sash
4. Verticals of frame
5. Horizontal frame and sill

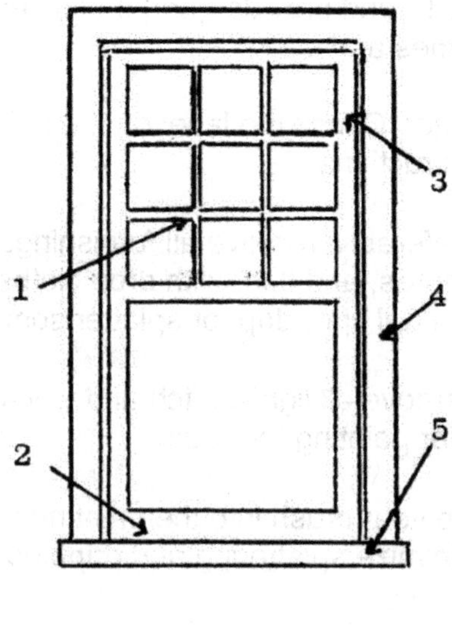

A simply way to protect the glass is to cover it with a piece of wet newspaper. The moisture will paste the newspaper to the glass and also prevent paint from soaking into the absorbent paper. When you strip the paper from the glass after painting, the paint will come with it.

CLEANUP

Brushes, rollers, and other equipment should be cleaned as soon as possible after use

Equipment used to apply latex paint can be easily cleaned with soap and water. Rinse thoroughly.

Equipment used to apply oil-based paint may be a little harder to clean. Soak brushes in turpentine or thinner long enough to loosen the paint. Then, work the bristles against the bottom of the container to release the paint. To release the paint in the center of the brush, squeeze or work the bristles between the thumb and forefinger. Rinse the brush in the turpentine or thinner again and, if necessary, wash it in mild soapsuds. Rinses in clear water.

After you clean a brush, comb the bristles to straighten them out, wrap the brush in heavy paper, and hang it up or store it flat. Never stand the brush on its bristles.

RELATED JOBS

Natural Finishes For Trim

Some doors, particularly flush doors, are attractive in their natural finish. However, they will discolor and soil easily unless protected. Your paint dealer can offer suggestions on how to finish and protect your doors. Many kinds of products are now on the market and new ones often appear.

The first step in finishing doors is to obtain the proper color tone. This is usually acquired by staining. However, sometimes no staining is required—the preservative finish is enough to bring out the desired color tone. With new doors, to help you make a decision, you can experiment on the trimmings or shavings.

The next step is sealing. One coat of shellac is usually adequate. When the shellac is dry, the surface should be sanded smooth, wiped free of dust, and varnished. Rubbing the surface with linseed oil, as is done in furniture finishing, provides a nice, soft finish, but requires more work. Also, surfaces so finished collect dust more readily.

For a natural finish of other interior trim, you need to specify the desired kind and grade of wood at the time of construction. This can add substantially to the construction costs.

Wood Floors

You may want to refinish your wood floors to complement your paint job. This should be done before you paint.

Complete renewal of the floors requires complete removal of the old finish. This can be done by sanding or with paint and varnish remover. Sanding is probably the fastest and easiest method. Electric sanders can be rented. Be sure to sand with the grain of the wood until you have a clean, smooth surface.

To retain the natural color, hardwood floors should be refinished with varnish or shellac. To change the color, stain may be applied—preferably on the raw wood. Oil stains are the easiest to work with.

One or more coats of wax will help protect your new floors.

Paint dealers generally have instruction pamphlets on re-doing floors.

Concrete Floors

Concrete floors can be painted, but it is important to use an enamel that has good alkali resistance. There are good rubber-based, epoxy, and urethane types available. Also available and recommended are latex paints made especially for concrete floors.

Clean dirt and grease from concrete floors before you paint them. Trisodium phosphate is a good cleaner to use.

Slick concrete floors should be roughened slightly before they are painted. To roughen or etch the floor, treat it with a solution of 1 gallon of muriatic acid mixed in 2 gallons of water. After treating, rinse the floor thoroughly and allow it to dry completely before you paint it.

FARM SERVICE BUILDING WALLS

Painting

Walls in farm-service buildings often must withstand almost constant rubbing by animals and frequent washings to remove manure and dirt. While durable paint is required, lead-based paint should not be used because the animals may lick the paint.

For these walls, use a catalyzed enamel—epoxy, polyester, or urethane type. Such paint costs more than ordinary paint, but it is more durable and washable. The ingredients usually come in two containers and must be mixed. Label instructions should be followed carefully for mixing and using the paint.

Whitewashing

Whitewashing is a relatively simple and inexpensive way to brighten the interior of livestock buildings. The whitewash may be applied with either a brush or a spray gun.

Surface Preparation

Removal all dirt, scale, and loose material by scraping or brushing with a wire brush. Many whitewashing jobs have been quite satisfactory without further surface preparation. However, for the best job, wash off as much of the old coat of whitewash as possible with hot water and vinegar or a *weak* hydrochloric acid solution.

Dampen the walls before applying whitewash. Unlike most paints, the application and adherence of whitewash are improved when the surface is slightly damp.

Mixing

Lime paste is the basis of whitewash. It may be prepared by either:

a. Soaking 50 pounds of hydrated lime in 6 gallons of water. (Refined limes such as chemical hydrate, agricultural spray hydrate, finishing lime, and pressure hydrated lime have fewer lumps and will make a smoother paste.)

b. Slaking 25 pounds of quick-lime in 10 gallons of boiling water. Cover and allow to slake at least 4 days.

Each of these preparations makes about 8 gallons of paste.

Different whitewash mixes are suggested for different surfaces. (Smaller batches of whitewash may be prepared by reducing the ingredients by an equal proportion in the formulas given below.)

For General Woodwork

Dissolve 15 pounds of salt in 5 gallons of water. Add this solution to the 8 gallons of paste, stirring constantly. Thin the preparation to the desired consistency with fresh water.

To reduce chalking, use 5 pounds of dry calcium chloride instead of the salt.

For Brick, Concrete, or Stone

Add 25 pounds of white Portland cement and 25 pounds of hydrated lime to 8 gallons of water. Mix thoroughly to a thick slurry. Thin to the consistency of thick cream. Mix only enough for a few hours' use.

To reduce chalking, add 1 to 2 pounds of dry calcium chloride dissolved in a small amount of water to the mix just before using.

For Plastered Walls

Two formulas are recommended:

a. Soak 5 pounds of casein in 2 gallons of water until thoroughly softened (about 2 hours). Dissolve 3 pounds of water softener (trisodium phosphate) in 1 gallon of water, add this solution to the casein, and allow the mixture to dissolve. When the lime paste and the casein are thoroughly cool, slowly add the casein solution to the lime, stirring constantly.

   Just before use, dissolve 3 points of formaldehyde in 3 gallons of water, and add this solution to the whitewash batch, stirring constantly and vigorously. *Do not add the formaldehyde too rapidly*. If the solution is added too fast, the casein may form a jelly-like mass, thus spoiling the batch.

b. Dissolve 3 pounds of animal glue in 2 gallons of water. Add this solution to the lime paste, stirring constantly. Thin the mixture to the desired consistency.

General-Use, Long-life Mix

The first formula, or mix, given for use on plastered walls, above, is a time-tested, long-life mix also suitable for general use. The following is also:

Dissolve 6 pounds of salt in 3 gallons of boiling water. Allow the solution to cool, and then add it to the lime paste. Stir 3 pounds of white Portland cement into the mix.

Coloring

Pigments may be added to whitewash to provide color. The following have proved satisfactory:

Black:	Magnetic black oxide of iron
Blue:	Ultramarine or cobalt blue
Brown:	Pure precipitated brown oxide of iron or mixtures of black oxide of iron with turkey or Indian red
Green:	Chromium oxide, opaque, or chromium oxide, hydrated
Red:	Indian red made from pure ferric oxide
Violet:	Cobalt violet and mixtures of reds, white, and blues
White:	Lime itself
Yellow:	Precipitated hydrated iron oxides

Application

Some surfaces may require two coats of whitewash. Two coats are better than one coat that is too heavy.

Strain the mix through three layers of cheesecloth before using a spray gun.

BASIC FUNDAMENTALS OF PAINTS AND PRESERVATIVES

TABLE OF CONTENTS

BASIC FUNDAMENTALS OF
PAINTS AND PRESERVATIVES

Painting is an expensive procedure, involving hours of time preparing the surface and applying the paint. The paint, itself, is expensive—not to mention the brushes, sprayers, respirators, and other associated equipment.

Every painter should see that each job of painting is done properly so that the best possible use is made of time, equipment, and material and so that it is not necessary to re-do a job before it would normally be required.

I. PURPOSES OF PAINTING

To employ paint materials and painting man-hours effectively and economically, the fundamental purposes of painting must be borne in mind. The importance of each of these purposes depends, of course, on the particular surface which is to be painted. Following is a brief discussion of each of these fundamental purposes of painting.

PREVENTIVE MAINTENANCE

The primary purpose of painting is protection. This is provided initially with new construction and maintained by a sound and progressive preventive maintenance program.

Resistance to moisture from rain, snow, ice, and condensation constitutes perhaps the greatest single protective characteristic of paint. All things made of metal corrode. Moisture causes wood to swell, warp, and rot. Interior wall finishes of buildings are ruined by neglect of exterior surfaces. Porous masonry is attacked and destroyed by moisture. Paint films must therefore be as impervious to moisture as possible to provide a protective waterproof film over the surface to which applied. Paint also acts as a protective film against attack by acids, alkalies, or marine organisms.

HABITABILITY

A compartment or room painted in pastel tints is more pleasant to live in than a room painted a brilliant red or orange. It can be readily seen that the function paint extends further than merely material protection.

Painting is used as a sanitary measure. A smooth, washable, painted surface which can be cleaned easily, helps produce a clean and healthful atmosphere. Therefore, a painted compartment or room is a healthier place to live in than one that is unpainted.

Another purpose of paint is to reflect light. Used in the interior, light-colored paints reflect and distribute both natural and artificial light, and thus help secure maximum efficiency from the lighting system. Correct illumination helps you to do your job better and easier.

IDENTIFICATION

Another purpose of colored paint is the identification of objects. Red is used to identify firefighting equipment; yellow means caution; green means safety.

II. PAINTS

The term PAINT is broadly applied to any mixture designed to be spread on a surface in liquid form and to "dry" to a thin, permanent surface coating. By general custom, however, the term PAINT is often restricted to materials containing pigments and designed to obscure the underlying substrata. Oil paint consists of pigments dispersed in a drying oil—usually linseed oil. ENAMELS are paints designed to resist scrubbing and washings. (Enamels are obtainable in flat, semi-gloss and gloss.)

VARNISH is distinguished from paint by the fact that it contains little or no pigment, and is not designed to obscure the surface to which it is applied. OIL varnishes are usually a combination of drying oil with a synthetic resin. When the resin is glyceryl phthalate, the varnish is referred to as an ALKYD VARNISH. SPIRIT VARNISHES are made by dissolving a resin, usually SHELLAC in alcohol.

LACQUERS, which may be clear or pigmented, consist of a cellulose derivative, commonly nitrocellulose, dissolved in a suitable solvent.

Paints which are applied to bare wood or metal surfaces to form undercoats for subsequent coats are called PRIMERS. The most common primers for metal surfaces contain anticorrosive pigments, such as RED LEAD, ZINC DUST, or ZINC CHROMATE. Usually primers for wood are specially formulated to adhere to the wood and to form a good surface for top coats. Paints designed to resist weather and sunlight are called EXTERIOR or OUTSIDE paints; paints not primarily so designed are called INTERIOR or INSIDE paints. Paints which dry to a dull finish are called FLAT paints; paints which dry to a shiny finish are called GLOSS paints; paints with an intermediate surface are called EGGSHELL or SEMI-GLOSS paints.

Paints are composed of various ingredients such as: PIGMENT; NONVOLATILE VEHICLE or BINDER; and SOLVENT or THINNER.

PIGMENTS are insoluble solids, divided finely enough to remain suspended in the vehicle for a considerable time after thorough stirring or shaking. There are several types of pigments.

OPAQUE pigments give the paint its hiding or covering capacity, and contribute other properties. The commonest opaque pigments are WHITE LEAD, ZINC OXIDE, and TITANIUM DIOXIDE.

COLOR pigments give the paint its color. This may be inorganic, such as CHROME GREEN, CHROME YELLOW, or IRON OXIDE; or ORGANIC, such as TOLUIDINE RED or PHTHALOCYANINE BLUE.

TRANSPARENT or EXTENDER pigments contribute bulk, and also control the application properties, durability, and resistance to abrasion of the coating.

MISCELLANEOUS pigments; there are many other pigments that are used for a variety of special purposes. Some of these pigments are; anticorrosive—metallic zinc dust; safety markings—luminous; and heat resistant—aluminum.

The VEHICLE, or BINDER, of paint is the material that holds the pigment together and also adheres to the surface. In general, the durability of the paint is determined by the resistance of the binder to the exposure conditions.

Formerly, linseed oil was the commonest binder, and it is still used in certain paints. It has however, largely been superseded by various synthetic resins. ALKYD resins are the commonest. These are made by the reaction of glyceryl phthalate and an oil, and may be made with almost any properties desired. Other common synthetic resins, which may be used by themselves or mixed with oil, include PHENOLICS, VINYLS, EPOXIES, URETHANES, POLYESTERS, CHLORINATED RUBBER, etc. Each has its own advantages and disadvantages. It is particularly important in the newer materials that the manufacturer's instructions be followed implicitly.

Certain synthetic materials, called LATEXES, are dispersed in water. Paints made from these are useful because they can be applied to damp surfaces, and tools and spills may be cleaned up easily with water. They have extremely high alkali resistance, and many have excellent durability. They are particularly useful for plaster and masonry surfaces. There are many different chemicals involved in latexes, but the commonest are STYRENE-BUTADIENE (or "synthetic rubber"), POLYVINYL ACETATE ("PVA" or "VINYL"), and ACRYLIC. All are very similar in their performance.

Other common binders are portland cement (in a dry-powder form to be mixed with water) and bituminous material (usually asphalt or coal-tar).

The only purpose of a SOLVENT is to adjust the consistency of the material so that it may be applied readily to the surface. The solvent then evaporates, contributing nothing further to the film. For this reason the cheapest SUITABLE solvent should be used. The solvents most used are MINERAL SPIRITS and NAPTHA. TURPENTINE is sometimes used, but contributes little that other solvents do not, and costs much more. Many synthetic resins require a special solvent, and it is IMPORTANT THAT THE CORRECT ONE BE USED, otherwise the paint may be entirely spoiled. Cement paints usually use water as a solvent.

III. STORAGE OF MATERIALS

Most paints, except cement-water paints, are ~~usually provided ready-mixed in 1-gal, 5-gal,~~ and 55-gal containers. Large quantities of paint in 1-gal and 5-gal containers should be stored in enclosures with fireproof walls; small quantities should be stored in properly constructed storage cabinets. Metal cabinets should be used if available; if not, cabinets should be constructed of asbestos-cement board not less than 5/32 in. thick. Bottoms and sides should be double thickness, with a 1 1/2-in. air space between the boards. Doors should also be of double thickness, with raised sills 2 in. above the bottoms of the cabinets. Doors should be provided with suitable locks, and a door should be kept closed and locked whenever paint is not being taken from or stored in the cabinet. All doors should be marked DANGER! FLAMMABLE! KEEP FLAME AND EXCESSIVE HEAT AWAY.

All mixed paint must be stored in nearly filled, tightly sealed containers, to prevent skinning over, the loss by evaporation of volatile materials, and the danger of fire.

Paint in storage should be arranged so that the oldest paint of each type is the first available. If old paint must be used with new paint, the entire lot should be blended to ensure uniform gloss and color.

The amount and kinds of EQUIPMENT available will depend on the shop to which you are assigned. This equipment may include either a paint spray outfit of 5-gallon capacity, or a lightweight, portable, 1-quart capacity sprayer, driven either by air or electrically. The equipment may also include a paint mixer of the type used with a portable electric or pneumatic drill, and respirators of the chemical-cartridge or mechanical-filter type. The number and types of brushes available also will vary.

You will have to use your own judgment as to the number of brushes to be kept available for daily use. The equipment should include a stencil-cutting machine with supplies and several sets of metal stenciling letters. (Incidentally, the paper used in stencil-cutting machines is referred to as stencilboard.)

This equipment must be properly maintained to prolong its life and to derive best results from its use. Before new paint brushes are used they should be rinsed with thinner. This tightens the bristles and also removes those which are loose. Brushes should not be soaked in water to tighten the bristles as this will cause the metal ferrule to rust or split due to the swelling of the wooden handle. Brushes that ~~are to be reused the following day should be~~ marked for white, light colors, or dark colors. Excess paint should be removed with thinner and the brushes suspended by the handle with the bristles immersed in thinner or linseed oil to just below the bottom of the ferrule. The weight of the brush must not rest on the bristles as that will cause them to become distorted. Brushes that are not to be reused immediately should be carefully cleaned with thinner, of the type recommended by the manufacturer, washed thoroughly with soap and water, then rinsed. A protective cover and preservative should be applied when appropriate. They should be stored suspended from racks or laid flat.

To clean a frozen brush, soak it in a solvent-type, nonflammable paint and varnish remover, squeeze and scrape the softened paint out of the bristles, and then clean the brush with thinner as previously described.

The spraying equipment used is of very high quality and will give excellent service for years if it is given proper care. Most frequent causes of unsatisfactory operation are faulty assembly, improper adjustment, and clogging because of dirt or hardened paint. Spray equipment should be cleaned with an appropriate thinner after each job or at the end of each day.

The paint supply hose should be disconnected from the tank and a container of thinner connected. Pulling the trigger will force the thinner through the paint hose and gun, which cleans out the paint remaining in them. The gun should be taken apart and each part cleaned. Care should be taken not to soak the packing or lubricated parts with thinner, as this will remove the lubricant and cause the packing to become hard. The paint tank should also be cleaned with thinner and wiped dry. All the equipment should be stored in its assigned place. The air and paint hoses should always be coiled before being stored.

The paint-mixing attachment should be removed from the electric or pneumatic drill and cleaned with thinner. The attachment should be removed prior to cleaning, because thinner will cause deterioration of the electric motor. Respirators used in spray-painting should be thoroughly cleaned with thinner after being used, to remove the accumulation of paint. They should then be wiped with a light soap and water solution to remove the thinner, wiped with clear fresh water, and thoroughly dried. If left damp, the metal parts will rust and the rubber will deteriorate.

It is advisable to wipe each respirator with a diluted disinfectant solution, since more than one individual may wear it. The filters or chemical cartridges should be removed and checked after each use and renewed when necessary. Do not use paint remover to clean the respirator, paint hose, mixer, or spray gun, as the corrosive agent contained in the remover will cause deterioration of this equipment.

IV. BRUSHES

Brushes, as any other tools, must be of first quality and maintained in perfect working condition at all times. Brushes are identified, first, by the type of bristle used. Brushes are made with either natural, synthetic or mixed bristles. Chinese hog bristles represent the finest of the natural bristles because of their length, durability and resiliency. Hog bristle has one unique characteristic in that the bristle end forks out like a tree branch. This "flagging" permits more paint to be carried on the brush and leaves finer brush marks on the applied coating which flow together more readily resulting in a smoother finish. Horsehair bristles are used in cheap brushes and are a very unsatisfactory substitute. The ends do not flag, the bristles quickly become limp, they hold far less paint and do not spread it as well. Brush marks left in the applied coating tend to be coarse and do not level out as smoothly. Some brushes contain a mixture of hog bristle and horsehair, and their quality depends upon the percentage of each type used. Animal hair is utilized in very fine brushes for special purposes. Badger hair, for example, produces a particularly good varnish brush. Squirrel and sable are ideal for striping, lining, lettering and free-hand art brushes. Of the synthetics, nylon is by far the most common. By artificially "exploding" the ends and kinking the fibres, manufacturers have increased the paint load nylon can carry, and have reduced the coarseness of brush marks. Nylon is steadily replacing hog bristle because of the difficulties in importing the latter. Nylon is almost always superior to horsehair. The very fact that nylon is a synthetic makes it unsuitable for applying lacquer, shellac, many creosote products and some other coatings that would soften or dissolve the bristles. Because water does not cause any appreciable swelling of nylon bristles, they are especially recommended for use with latex paints. Brushes are further identified by types, that is, the variety of shapes and sizes as are required for specific painting jobs. Types can be classified as follows: (See figs. 1 and 2.)

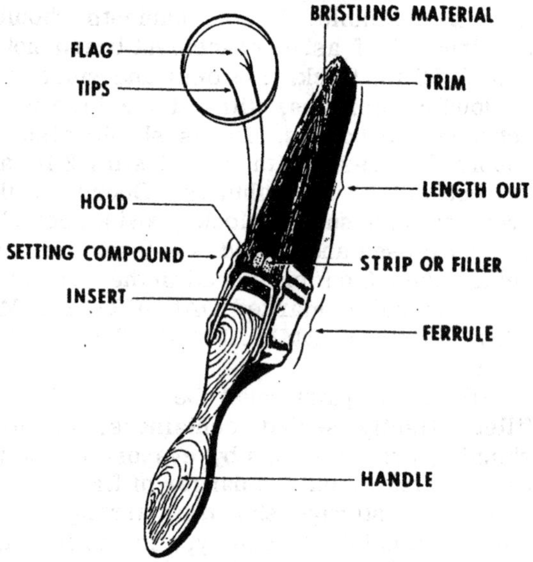

Figure 1.—Typical paint brush.

1. WALL BRUSHES: Flat, square-edged brushes ranging in widths from 3" to 6" and used for painting large, continuous surfaces, either interior or exterior.

2. SASH AND TRIM BRUSHES: Available in four shapes, flat square-edged, flat angle-edged, round and oval. These brushes range in width from 1 1/2" to 3" or diameters of 1/2" to 2" and are used for painting window frames, sash, narrow boards, also interior and exterior trim surfaces. For fine-line painting, the edge of the brush is often chisel-shaped to make precise edging easier to accomplish.

3. ENAMELING AND VARNISH BRUSHES: Flat square-edged or chisel-edged brushes available in widths from 2" to 3". The select, fine bristles are comparatively shorter in length to cause relatively high viscosity gloss finishes to lay down in a smooth, even film.

4. STUCCO and MASONRY BRUSHES: These have the general appearance of flat wall brushes and are available in widths ranging from 5" to 6". Bristles can be of hog, other

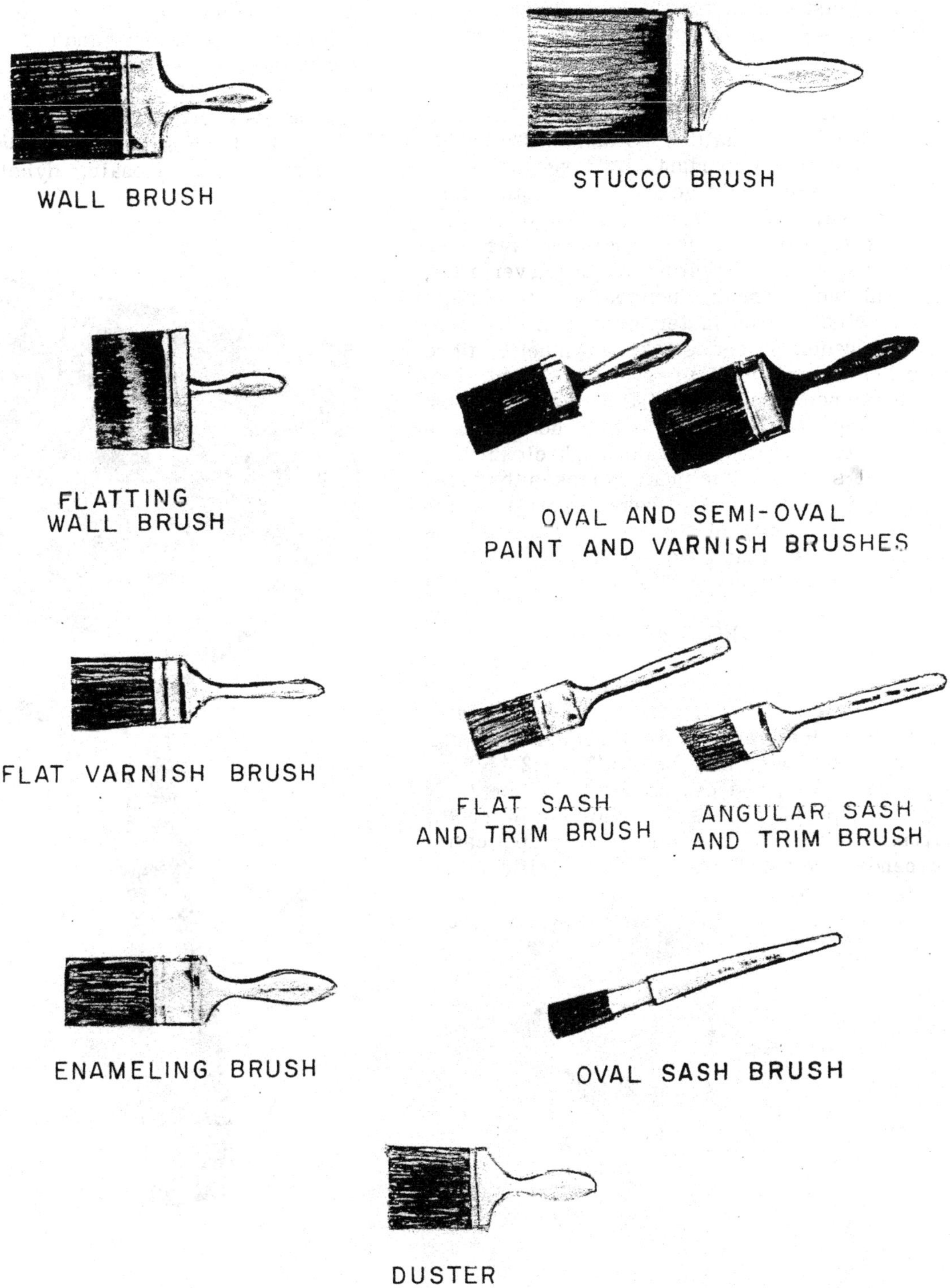

Figure 2.—Types of brushes.

natural bristle or nylon; the latter is preferred for rough surfaces because of its resistance to abrasion.

Use the right size brush for the job. Avoid a brush that is too small or too large. The latter is particularly important. A large-area job does not necessarily go faster with an oversize brush. If the brush size is out of balance for the type of painting being done, the user tends to apply the coating at an uneven rate, general workmanship declines, and the applicator actually tires faster because of the extra output required per stroke. Synthetic fibre brushes are ready to use when received. The performance of natural bristle brushes is very much improved by a previous 48 hour soak in linseed oil followed by a thorough cleaning in mineral spirits. This process makes the bristles more flexible and serves to swell the bristles in the ferrule of the brush resulting in a better grip so that fewer bristles are apt to work loose when the brush is used.

V. ROLLERS

A paint roller consists of a cylindrical sleeve or cover which slips on to a rotatable cage to which a handle is attached. (See fig. 3.) The cover may be 1 1/2" to 2 1/4" inside diameter, and usually 3", 4", 7" and 9" in length. Special rollers are available in lengths from 1 1/2" to 18". Proper roller application depends on the selection of the specific fabric

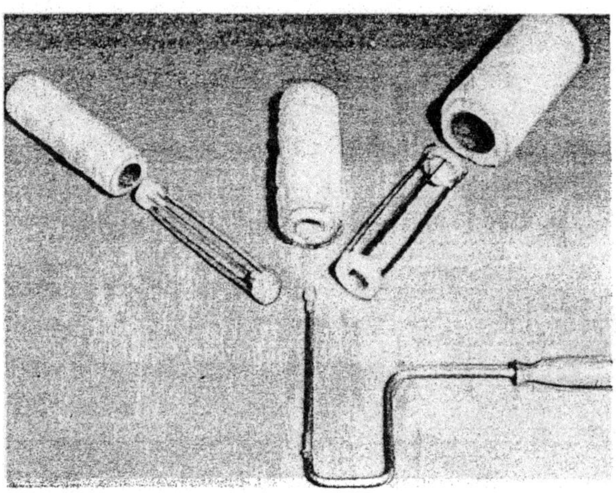

Figure 3.—Parts of a roller

and the thickness of fabric (nap length) based on the type of paint used and the smoothness or roughness of the surface to be painted. Special rollers are used for painting pipes, fences and other hard-to-reach areas. (See figs. 4 and 5.) The fabrics generally used for rollers are lambs wool, mohair, dynel, dacron and rayon.

Figure 4.—Pipe roller.

LAMBS WOOL (pelt) is the most solvent resistant type of material used and is available in nap lengths up to 1 1/4". It is recommended for synthetic finishes for application on

Figure 5.—Fence roller.

semi-smooth and rough surfaces. It mats badly in water, and is not recommended for water paints.

MOHAIR is made primarily of Angora hair. It also is solvent resistant and is supplied in 3/16 and 1/4 inch nap length. It is recommended for synthetic enamels and for use on smooth surfaces. It can be used with water paints also.

DYNEL is a modified acrylic fibre which has excellent resistance to water. It is best for application of conventional water paints and solvent paints, except those which contain strong solvents, such as ketones. It is available in all nap lengths from 1/4" to 1 1/4".

DACRON is a synthetic fibre which is somewhat softer than DYNEL. It is best suited for exterior oil or latex paints. It is available in nap lengths from 5/16" to 1/2".

RAYON fabric is not recommended because of the poor results generally obtained from its use. Furthermore, rayon mats badly in water. Table 1 can be used as a guide for choosing the proper roller cover.

Immediately after use, rollers should be cleaned with the type of thinner recommended for the paint in which the roller was used. After cleaning with thinner, the roller should be thoroughly washed in soap and water, rinsed in clear water, and dried.

VI. SPRAY GUNS

A spray gun is a precision tool that mixes air under pressure with paint, breaks it up into spray, and ejects it out in a controlled pattern.

There are several types, either with a container attached to the gun or with the gun connected to a separate container by means of hoses. There are bleeder or non-bleeder, external- or internal-mix, and pressure-, gravity-, or suction-feed guns.

The BLEEDER type of gun is one in which air is allowed to leak—or bleed—from some part of the gun in order to prevent air pressure from building up in the air hose. In this type of gun the trigger controls the fluid only. It is generally used with small air compressing outfits that have no pressure control on the air line.

The NONBLEEDER gun is equipped with an air valve which shuts off the air when the trigger is released. It is used with compressing outfits having a pressure-controlling device.

An EXTERNAL-MIX gun is one which mixes air and paint outside and in front of the gun's air cap. This type of gun can do a wide variety of work and has the power to throw a very fine spray, even of heavy material. It also permits exact control over the spray pattern. An external-mix air cap is shown in figure 6.

An INTERNAL-MIX spray gun mixes the air and fluid inside the air cap as pictured in figure 7. It is not as widely used as the external-mix gun.

In a SUCTION-FEED spray gun, the air cap, shown in figure 8, is designed to draw the fluid from the container by suction—in somewhat the same way that an insect spray gun operates. The suction-feed spray gun is usually used with 1-quart (or smaller) containers.

A PRESSURE-FEED gun operates by air pressure, which forces the fluid from the container into the gun. This is the type (fig. 9) used for large-scale painting.

Table 1.–Roller Selection Guide

Type of Paint	*Smooth (1)*	*Type of Surface* *Semi-smooth (2)*	*Rough (3)*
Aluminum	C	A	A
Enamel or Semigloss (Alkyd)	A or B	A	
Enamel undercoat	A or B	A	
Epoxy coatings	B or D	D	D
Exterior House Paint:			
Latex for wood	C	A	
Latex for masonry	A	A	A
Oil or alkyd—wood	C	A	
Oil or alkyd—masonry	A	A	A
Floor enamel—all types	A or B	A	
Interior Wall paint:			
Alkyd or oil	A	A or D	A
Latex	A	A	A
Masonry sealer	B	A or D	A or D
Metal primers	A	A or D	
Varnish—all types	A or B		
*Roller Cover Key**		*Nap Length* (inches)	
A—Dynel (modified acrylic)	$\frac{1}{4}$–$\frac{3}{8}$	$\frac{3}{8}$–$\frac{3}{4}$	1–1$\frac{1}{4}$
B—Mohair	$\frac{3}{16}$–$\frac{1}{4}$		
C—Dacron polyester	$\frac{1}{4}$–$\frac{3}{8}$	$\frac{1}{2}$	
D—Lambswool pelt	$\frac{1}{4}$–$\frac{3}{8}$	$\frac{1}{2}$–$\frac{3}{4}$	1–1$\frac{1}{4}$

(1) Smooth Surface: hardboard, smooth metal, smooth plaster, drywall, etc.

(2) Semi-smooth Surface: sand finished plaster and drywall, light stucco, blasted metal, semi-smooth masonry.

(3) Rough Surface: concrete or cinder block, brick, heavy stucco, wire fence.

* Comprehensive product standards do not exist in the Paint Roller Industry. Roller covers vary significantly in performance between manufacturers and most manufacturers have more than one quality level in the same generic class. This table is based on field experience with first line products of one manufacturer.

PARTS OF THE SPRAY GUN

The two main assemblies of the spray gun are the gun body assembly and the spray head assembly. Each of these assemblies is a collection of small parts, all of which are designed to do specific jobs.

The principal parts of the gun body assembly are shown in figure 10. The air valve controls the supply of air and is operated by the trigger. The spreader adjustment valve regulates the amount of air that is supplied to the spreader horn holes of the air cap, thus varying the paint pattern. It is fitted with a dial which can be set to give the pattern desired. The fluid needle adjustment controls the amount of spray material that passes through the gun. The spray head locking bolt locks the gun body and the removable spray head together.

Most guns are now fitted with a removable spray head assembly. This type has many advantages. It can be cleaned more easily, it permits quick change of the head when you want to use a new color or material, and, if it is damaged, a new head can be put on the old gun body.

The principal parts of the spray head assembly are the air cap, the fluid tip, fluid needles, and spray head barrel, pictured in figure 11.

The fluid tip regulates the flow of the spray material into the air stream. The tip encloses the end of the fluid needle. The spray head barrel is the housing which encloses the head mechanism.

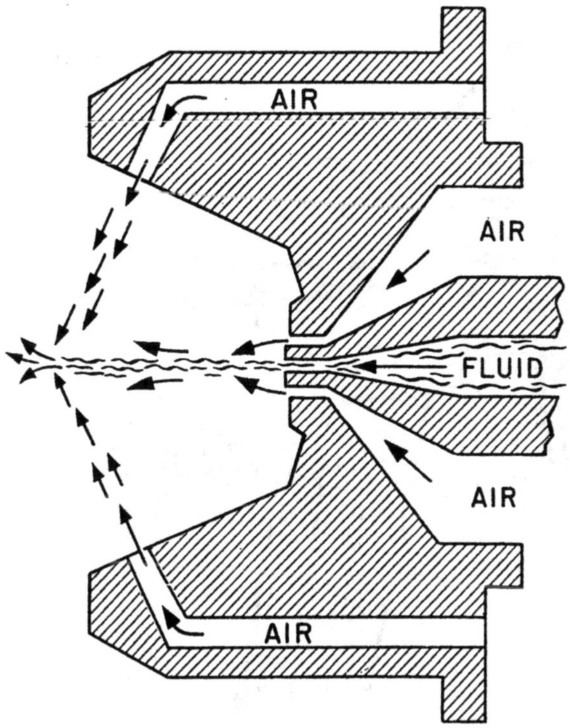

Figure 6.—An external-mix air cap.

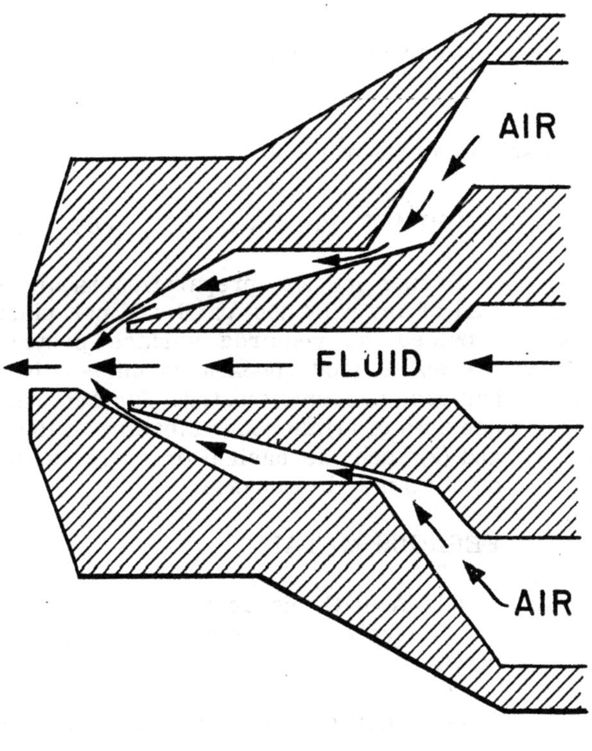

Figure 7.—An internal-mix air cap.

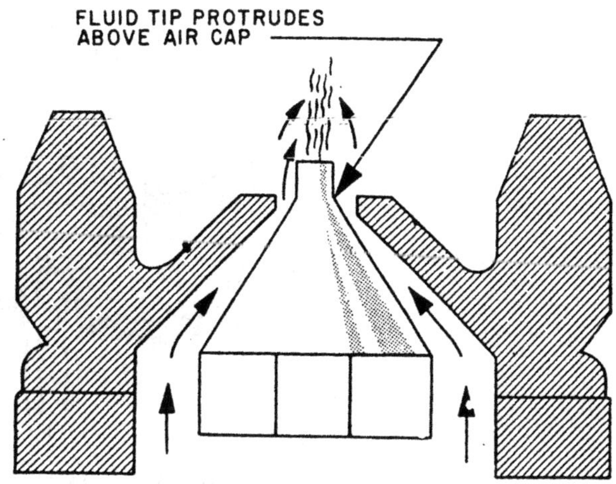

Figure 8.—A suction-feed air cap.

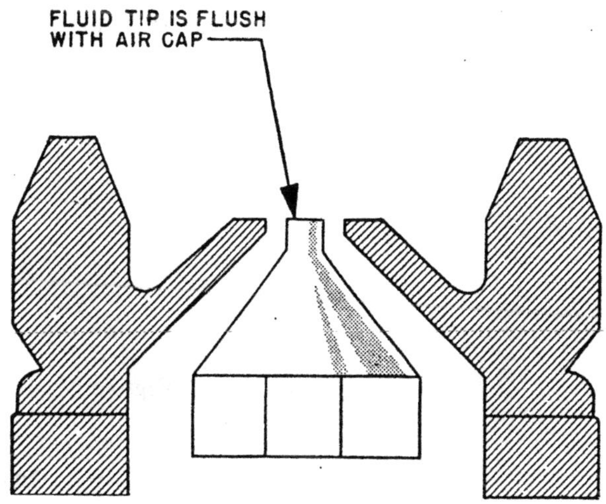

Figure 9.—A pressure-feed air cap.

LUBRICATION OF THE SPRAY GUN

Your spray gun also needs lubrication. The fluid needle packing should be removed occasionally and softened with oil. The fluid needle spring should be coated with grease or petrolatum. Figure 12 shows where these parts are and also the oil holes in which you occasionally should put a few drops of light oil.

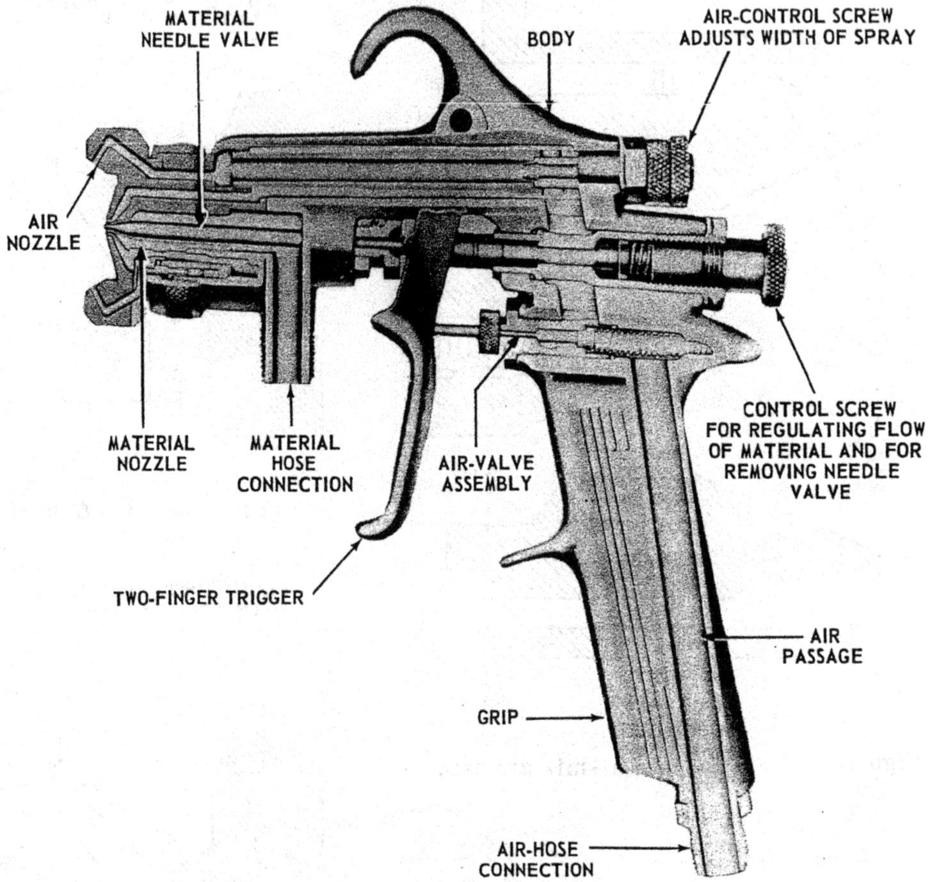

Figure 10.—Cross section of a spray gun.

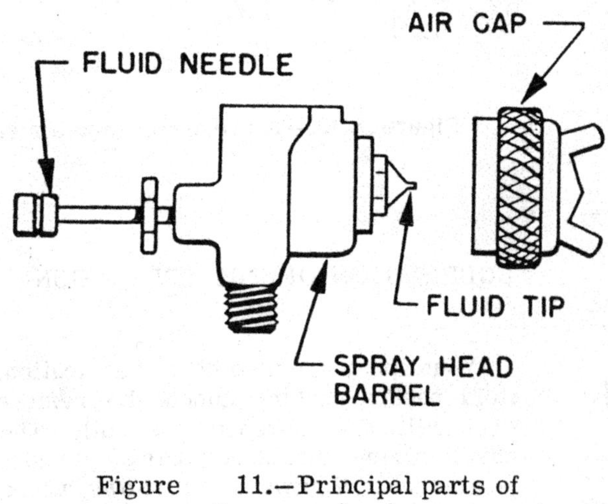

Figure 11.—Principal parts of the spray head.

SURFACE PREPARATION FOR PAINTING

Proper surface preparation is an essential part of any paint job; paint will not adhere well, provided the required surface protection, or present a good appearance unless the surface has been properly treated. Surface preparation consists of (1) thorough cleaning of the surface, and (2) such mechanical or chemical pretreatment as may be necessary.

FERROUS METALS

CLEANING the ferrous metals (iron and steel) involves the removal of oil, grease, and dirt .

This specification describes three types of

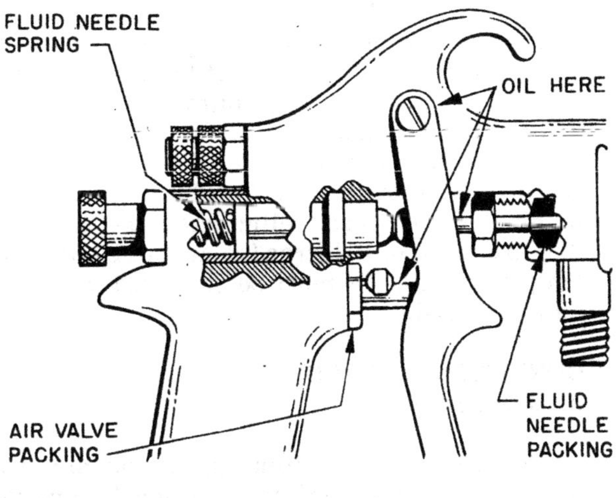

Figure 12.—Lubrication points of a spray gun.

ferrous-metal cleaners: Type 2, a HOT ALKALINE cleaner; Type 3, a SOLVENT cleaner, and Type 4, an EMULSION cleaner.

PRETREATMENT of the ferrous metals involves the application of a pretreatment coating.

If a red lead or zinc chromate primer is to be used, however, the pretreatment coating is omitted.

NONFERROUS METALS

The nonferrous metals are brass, bronze, copper, tin, zinc, aluminum, nickel, and other metals which are not derived from iron ore. A galvanized metal is one which has a zinc-coated surface; galvanized metals are therefore treated as non-ferrous metals for painting purposes.

Nonferrous metals are cleaned with a solvent-type cleaner and pretreated by coating with pretreatment coating.

CONCRETE AND MASONRY

Cleaning concrete and masonry involves the removal of dirt, mildew and EFFLORESCENCE (a white, powdery, crystalline deposit which often forms on concrete and masonry surfaces).

Dirt and fungus are removed by washing with a solution of TRISODIUM PHOSPHATE; the strength of the solution may vary from 2 oz to 8 oz of trisodium phosphate per gal of water, depending upon the amount of dirt and/or mildew on the surface. Immediately after washing, rinse off all the trisodium phosphate with clear water. If oil paint is to be used, allow the surface to dry thoroughly before painting.

For efflorescence, first remove as much as possible of the deposit by dry-brushing with a wire brush or a stiff fiber brush. Next, wet the surface thoroughly with clear water, and then scrub with a stiff brush dipped in a 5 percent solution (by weight) of muriatic acid. Allow the acid solution to remain on the surface about 3 minutes before scrubbing, but rinse thoroughly with clear water IMMEDIATELY AFTER scrubbing. Work on small areas, not larger than 4 sq ft in size. Wear rubber gloves, a rubber apron, and goggles when mixing and using the acid solution. In mixing the acid, ALWAYS add acid to water, rather than water to acid.

For a very heavy deposit, the acid solution may be increased to 10 percent and it may be allowed to remain on the surface for 5 minutes before scrubbing.

All defects in a concrete or masonry surface must be repaired before painting. To repair a large crack, cut the crack out to an inverted-V shape and plug it with grout made by mixing 2 or 3 parts of mortar sand, 1 part of portland cement, and enough water to make a putty-like consistency. After the grout sets, damp-cure it by keeping it wet for 48 hours. If oil paint is to be used, allow at least 90 days for weathering before painting over a grout-filled crack.

PLASTER AND WALLBOARD

Whenever possible, new plaster should be aged at least 2 months before painting. Prior to painting, fill all holes and cracks with SPACKLING COMPOUND or PATCHING PLASTER. Cut out the material along the crack or hole in inverted-V shape. To avoid excessive absorption of water from the patching material, wet the edges and bottom of the crack or hole before applying the material. Fill the opening to within 1/4 in. of the surface and allow the material partially to set before you bring the level up flush with the surface. After the material has thoroughly set, smooth it up with fine sandpaper. Allow at least 72 hours for setting before painting. Plaster and wallboard are primed with PRIMER

WOOD

Prior to painting, a wood surface should be closely inspected for loose boards, defective lumber, protruding nailheads, or any other defects or irregularities. Loose boards should be nailed tight, defective lumber should be replaced, and all nailheads should be countersunk.

A dirty wood surface is cleaned for painting by sweeping, dusting, and washing with solvent or soap and water. When washing wood, take care to avoid excessive wetting, which tends to raise the grain. Wash a small area at a time, and rinse and dry immediately.

Wood which is to be given a NATURAL finish (meaning wood which will not be concealed by an opague surface coating) may require BLEACHING to a uniform and/or light color. Bleaching is done by applying a solution of 1 lb of OXALIC ACID to 1 gal of hot water. More than one application may be required. After the solution has dried, smooth the surface with fine sandpaper.

Rough wood surfaces must be sanded smooth for painting. Mechanical SANDERS of various types are used for large areas. Hand-sanding of small areas is done by wrapping the sandpaper around a rubber, wood, or metal SANDING BLOCK. For a very rough surface, start with a coarse paper, about No. 2 or 2 1/2; follow up with a No. 1/2, No. 1, or No. 1 1/2; and finish with about a No. 2/0 grit. For fine work, such as furniture work, finish with a still finer grit.

Sap or resin in wood will stain through a coat, or even several coats, of paint. Remove sap or resin by scraping and/or sanding.

PAINT REMOVERS

Paint and varnish removers generally are used for small areas. Solvent type removers or solvent mixtures are selected according to the type and condition of the old finish as well as the nature of the substrate. Removers are available as flammable or non-flammable types, also liquid or semi-paste in consistency. While most paint removers require scraping or steel wool to physically remove the softened paint, types are available that allow the loosened finish to be flushed off with steam or hot water. Many of the flammable and non-flammable removers contain paraffin wax to retard evaporation. It is absolutely essential that this residue be removed from the surface prior to painting to prevent loss of adhesion of the applied coating. In such instances, follow the manufacturer's label directions or use mineral spirits to remove any wax residue. As a safety precaution, it should be noted that, while non-flammable removers eliminate fire hazards, they are toxic to a degree (as are all removers). Proper ventilation must be provided whenever they are used.

CONDITIONS, SEALERS, AND FILLERS

Conditioners are often applied on masonry to seal a chalky surface in order to improve adhesion of water-based topcoats. Sealers are used on wood to prevent resin exudation or bleeding. Fillers are used to produce a smooth finish on open grain wood and rough masonry. (See table 2.)

Latex (water-thinned) paints do not adhere well to chalky masonry surfaces. To overcome this problem, an oil-based CONDITIONER is applied to the chalky substrate before the latex paint is applied. The entire surface should be vigorously wire brushed by hand or power tools, then dusted to remove all loose particles and chalk residue. The conditioner is then brushed on freely to assure effective penetration and allowed to dry. This surface conditioner is not intended for use as a finish coat.

SEALERS are used on bare wood to prevent resin exudation (bleeding) through applied paint coatings. Freshly exuded resin, while still soft, may be scraped off with a putty knife and the affected area solvent cleaned with alcohol. Hardened resin may be removed by scraping or sanding. Since the sealer is not intended for use as a priming coat, it should be used only when necessary, and applied only over the affected area. When previous paint on pine lumber has become discolored over knots, the sealer should be applied over the old paint before the new paint is applied.

FILLERS are used on porous wood, concrete, and masonry to fill the pores to provide a smoother finish coat.

Wood fillers are used on open-grained hardwoods. In general those hardwoods with pores larger than in birch should be filled. (See table 3.) When filling is necessary, it is done after any staining operations. Stain should be allowed to dry for 24 hours before filler is applied. If staining is not warranted, natural

Table 2.—Treatment of Various Substrates

	Wood	Metal: Steel	Metal: Other	Concrete Masonry	Plaster Wallboard
Mechanical					
Hand Cleaning	S	S	S	S	S
Power Tool Cleaning	S*	S	......	S	......
Flame Cleaning	......	S	......	......	......
Blast Cleaning:					
Brush-Off	......	S	S	S	......
All Other	......	S	......	......	......
Chemical and Solvent					
Solvent Cleaning	S	S	S	......	......
Alkali Cleaning	......	S	......	S	......
Steam Cleaning	......	S	......	S	......
Acid Cleaning	......	S	......	S	......
Pickling	......	S	......	......	......
Pretreatments					
Hot Phosphate	......	S	......	......	......
Cold Phosphate	......	S	......	......	......
Wash Primers	......	S	S	......	......
Conditioners, Sealers and Fillers					
Conditioners	......	......	......	S	......
Sealers	S	......	......	......	......
Fillers	S	......	......	S	......

S—Satisfactory for use as indicated
*—Sanding only

(uncolored)filler is applied directly to the bare wood. The filler may be colored with some of the stain in order to accentuate the grain pattern of the wood. To apply, first thin the filler with mineral spirits to a creamy consistency, then liberally brush it across the grain, followed by light brushing along the grain. Allow to stand five to ten minutes until most of the thinner has evaporated, at which time the finish will have lost its glossy appearance. Before it has a chance to set and harden, wipe the filler off ACROSS the grain using burlap or other coarse cloth, rubbing the filler into the pores of the wood while removing the excess. Finish by stroking along the grain with clean rags. It is essential that all excess filler be removed. Knowing when to start wiping is important; wiping too soon will pull the filler out of the pores, while allowing the filler to set too long will make it very difficult to wipe off. A simple test for dryness consists of rubbing a finger across the surface. If a ball is formed, it is time to wipe. If the filler slips under the pressure of the finger, it is still too wet for wiping. Allow the filler to dry for 24 hours before applying finish coats.

Masonry fillers are applied by brush to bare and previously prepared (all loose, powdery, flaking material removed) rough concrete, concrete block, stucco or other masonry surfaces, both new and old. The purpose is to fill the open pores by brushing the filler into the surface to produce a fairly smooth finish. If the voids on the surface are large, it is preferable to apply two coats of filler, rather than one heavy coat, in order to avoid mud-cracking. Allow 1 to 2 hours drying between coats. Allow the final coat to dry for 24 hours before painting.

WEATHER AND TEMPERATURE

Oil-painting and water-painting should not be done in temperatures above 95° or below 45°. Varnishing, shellacking, lacquering, and enameling should not be done in temperatures below 65° or above 95°. No painting except water-painting should be done on a damp surface, or on one which is exposed to hot sunlight.

VII. PAINT MIXING AND CONDITIONING

Most paints used in the shop are READY-MIXED, meaning that most shop paints are

Table 3.—Characteristics of Wood

Type	*Soft*	*Hard*		
Grain	*Closed*	*Open*	*Closed*	*Notes on Finishing*
Name of Wood				
Ash	......	X	......	Requires filler.
Alder	X	......	......	Stains well.
Aspen	......	......	X	Paints well.
Basswood	......	......	X	Paints well.
Beech	......	......	X	Paints poorly; varnishes well.
Birch	......	......	X	Paints and varnishes well.
Cedar	X	......	......	Paints and varnishes well.
Cherry	......	......	X	Varnishes well.
Chestnut	......	X	......	Requires filler; paints poorly.
Cottonwood	......	......	X	Paints well.
Cypress	......	......	X	Paints and varnishes well.
Elm	......	X	......	Requires filler; paints poorly.
Fir	X	......	......	Paints poorly.
Gum	......	......	X	Varnishes well.
Hemlock	X	......	......	Paints fairly well.
Hickory	......	X	......	Requires filler.
Mahogany	......	X	......	Requires filler.
Maple	......	......	X	Varnishes well.
Oak	......	X	......	Requires filler.
Pine	X	......	......	Variable depending on grain.
Teak	......	X	......	Requires filler.
Walnut	......	X	......	Requires filler.
Redwood	X	......	......	Paints well.

Note: Any type finish may be applied unless otherwise specified.

provided with the ingredients already mixed together in the proper proportions. When oil paints are left in storage for a long while, however, the pigments settle to the bottom, and must be again mixed into the vehicle before the paint is used. This procedure is what is meant by the term "mixing" as used in this section.

MIXING TECHNIQUES

Whenever possible, mix paint in the paint shop. The shop is usually equipped with a mechanical AGITATOR which mixes paint by rapidly shaking the container. In the absence of an agitator, use a strong, smooth, clean wood or metal paddle. If the pigment has settled in a cake, pour the vehicle off into another container and break up the pigment with the paddle. Then pour the vehicle back in, a little at a time, while continuing to work in the pigment. Then BOX the paint by pouring it back and forth from one container to the other. Continue boxing until the pigment and vehicle form a smooth mixture of uniform consistency and color.

A newly-opened can of ready-mixed paint is usually of the proper consistency for applying. Eventually, however, the paint will thicken as the volatile portion of the thinner evaporates from the open can. When this happens, enough of the appropriate thinner must be added to bring the paint back to working consistency.

The same applies to the drier. When the paint takes longer than it should to dry, the drier has evaporated below the required level, and more drier should be added. Great care must be taken against adding too much drier, however. Paint containing too much drier will dry too rapidly on the surface, which may cause WRINKLING.

Oil paint should be stirred frequently during use, to keep the pigment from settling to the bottom. Varnish and shellac, however, should not be stirred or agitated. Enamel should be mixed with a hand-paddle, not with a shake-type mechanical agitator. A shake-type agitator whips air into enamel, causing it to bubble or froth. Bubbled or frothed enamel must be allowed to stand 6 to 8 hours before it can be used.

CONDITIONING PAINT

When a partially filled can of oil paint is placed in storage, the surface of the paint should be covered with a 1/16-in. layer of the appropriate thinner and the can should be covered as tightly as possible. The layer of thinner will prevent the paint from skinning over, and the tight cover on the can will prevent the thinner from evaporating.

To remove lumps, pieces of skin, or foreign materials from paint, strain the paint through a sieve made of fine wire mesh, silk, or cheesecloth. All paint used in spray guns must be thoroughly strained.

VIII. METHODS OF APPLYING PAINT

The most common methods of applying paint are by brush, roller and spray. Dip and flow coat methods are also used but the mechanics of application limit their use to shop work. Of the three designed for field use, brushing is the slowest method, rolling is much faster, and spraying is usually the fastest of all. The choice of method is based on many additional factors such as environment, type of substrate, type of coating to be applied, appearance of finish desired and skill of personnel involved in the operation.

The general surroundings may prohibit the use of spray application because of possible fire hazards or potential damage from overspray. Typical of these are parking lots and open storage areas. Adjacent areas, not to be coated, must be masked when spraying is performed. This results in loss of time and, if extensive, may offset the advantage of the rapidity of spraying operations.

Roller coating is most efficient on large flat surfaces. Corners, edges and odd shapes, however, must be brushed. Spraying also is most suitable for large surfaces, except that it can also be used for round or irregular shapes. Brushing is ideal for small surfaces or for cutting in corners and edges. Dip and flow coat methods are suitable for volume production painting of small items in the shop.

Rapid drying, lacquer type products, e.g., vinyls, should be sprayed. Application of such products by brush or roller may be extremely difficult especially in warm weather or outdoors on breezy days.

Coatings applied by brush may leave brush marks in the dried film; rolling leaves a stippled effect, while spraying yields the smoothest finish, if done properly.

Personnel require the least amount of training to use rollers, and the most training to use spray equipment. The degree of training and experience of personnel will influence the selection of the application method.

BRUSH AND PAINT APPLICATION

Select the type of brush and paint pot needed for the job. The best type of paint pot for brush painting is a 1-gallon paint can from which the lip around the top has been removed. (The lid of the can is fitted to the lip around the top.) you can cut this lip off with a cold chisel. If you leave the lip on the pot, it will fill up with paint as you scrape the brush, and this paint will be continually streaking down the outside of the pot and dripping off.

Dip the brush to only one-third the length of the bristles, and scrape the surplus paint off the lower face of the brush, so there will be no drip as you transfer the brush from the pot to the work.

Here is how to apply paint by brush. For complete coverage, follow the style and first "lay on," then "lay off." Laying on means applying the paint first in long, horizontal strokes. Laying off means crossing your first strokes by working up and down. (See fig. 13.)

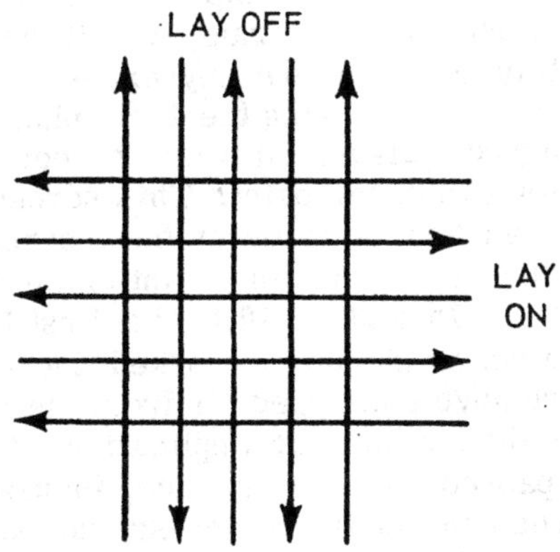

Figure 13.—Laying on and laying off.

By using the laying on and laying off method and crossing your strokes, the paint is distributed evenly over the surface, the surface is completely covered, and a minimum amount of paint is used. A good rule is to "lay on" the paint the shortest distance across the area and "lay off" the longest distance. When painting walls, or any vertical surface, "lay on" in horizontal strokes, "lay off" vertically.

Always paint ceiling first and work from the far corner. By working the ceiling first, you can keep the wall free of drippings by wiping up as you go along.

When painting ceiling surfaces, you will find that paint coats on the ceiling should normally be "lay on" for the shortest ceiling distance and "lay off" for the longest ceiling distance.

To avoid brush marks when finishing up a square, use strokes directed toward the last square finished, gradually lifting the brush near the end of the stroke while the brush is still in motion. Every time the brush touches the painted surface at the start of a stroke, it leaves a mark. For this reason, never finish a square by brushing toward the unpainted area, but always end up by brushing back toward the area already painted.

When painting pipes and stanchions and narrow straps, beams, and angles, lay the paint on diagonally as shown in figure 14. Lay off along the long dimension.

Always carry a rag for wiping dripped or smeared paint.

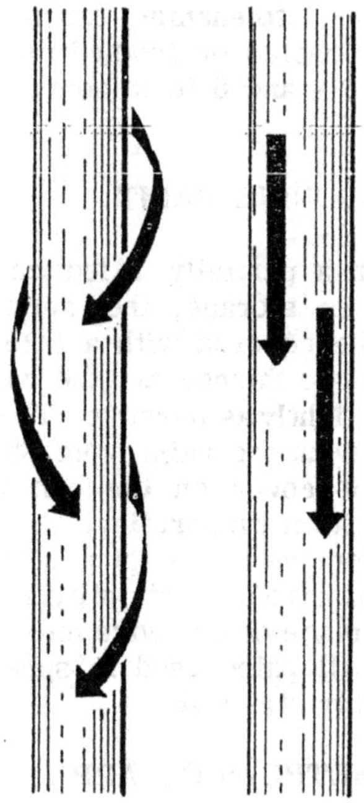

Figure 14.—Painting pipes and stanchions.

ROLLER METHOD

Pour the pre-mixed paint into the tray to about one-half of the depth of the tray. Immerse the roller completely, then roll it back and forth along the ramp to fill the cover completely and remove any excess paint. As an alternative to using the tray, place a specially designed galvanized wire screen into a five gallon can of the paint. This screen attaches to the can and remains at the correct angle for loading and spreading paint on the roller. (See figs. 15 and 16.) The first load of paint on a roller should be worked out on newspaper to remove entrapped air from the roller cover. It is then ready for application. As the roller is passed over a surface, thousands of tiny fibres continually compress and expand, metering out the coating and wetting the surface. This is in sharp contrast to other application methods that depends upon the skill and technique

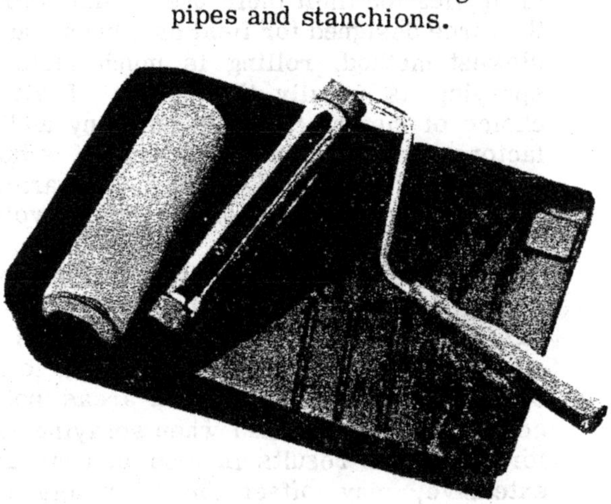

Figure 15.—Roller and tray.

of the painter. The uniformity of application by roller is less susceptible to variance in painter ability than other methods. Basic rules must still be followed. Always trim around ceilings, moldings, etc., before rolling the major wall or ceiling surfaces. Then roll as close as possible to maintain the same texture. Trimming is

Figure 16.—Roller and wire screen attachment to can.

usually done with a 3 inch wall brush. Always roll paint onto the surface, working from the dry area into the just-painted area. Never roll completely in the same or one direction. Don't roll too fast and avoid spinning the roller at the end of the stroke. Always feather out final strokes to pick up any excess paint on the surface. This is accomplished by rolling the final stroke out with minimal pressure.

SPRAY METHOD

Complete instructions for the care, maintenance, and operation of a spray gun are contained in the manufacturer's manual, and these instructions should be carefully followed. Only a few of the major spray-painting techniques can be given here, as follows:

SPRAY GUN ADJUSTMENT

The first essential is the correct adjustment of the AIR CONTROL and MATERIAL CONTROL screws, to produce the type of spray best suited to the nature of the work. The air control screw adjusts the width and the density of the spray. Turning the screw clockwise concentrates the material into a round, more dense spray; turning it counterclockwise widens the spray into a fan-shaped, more diffused spray. As the spray is widened, the flow of material must be increased; if it is not, the spray will break into a fog. Turning the material control screw clockwise increases the flow of material; turning it counterclockwise decreases the flow. The most desirable character of spray (from round and solid to fan-shaped and diffused) depends upon the character of the surface and the type of material being sprayed. Experience and experiment are about the only guides here. Practice spraying should be done on waste material, using different practice adjustments, until a spray is obtained which covers uniformly and adequately.

OPERATIONAL DEFECTS OF THE SPRAY GUN

Uneven distribution of the spray pattern is caused by clogging of one or more of the air outlets or by incorrect adjustment of the air and/or material controls.

SPITTING is the alternate discharge of paint and air. Common causes of spitting are drying of the packing around the material control needle valve, looseness of the material nozzle, and dirt in the material nozzle seat. To remedy dry packing, back off the material control needle valve and place two drops of machine oil on the packing. To remedy looseness of the material nozzle and dirt on the nozzle seat, remove the nozzle, clean the nozzle and seat with thinner, and screw the nozzle tightly back into place.

AIR LEAKAGE from the front of the gun is usually caused by improper seating of the air valve in the AIR VALVE ASSEMBLY shown in figure 10. Improper seating may be caused by foreign matter on the valve or seat, by wear on or damage to the valve or seat, by a broken valve spring, or by sticking of the valve stem caused by lack of lubrication.

PAINT LEAKAGE from the front of the gun is usually caused by improper seating of the material needle valve. Improper seating may be caused by damage to the valve stem or tip, by foreign matter on the tip or seat, or by a broken valve spring.

SPRAY-GUN STROKE

Figure 17 shows the correct method of stroking with a spray gun. Hold the gun 6 to 8 in. from the surface to be painted, keep the axis of the spray perpendicular to the surface, and take strokes back and forth in horizontal lines. Pull the trigger just after you start a stroke, to avoid applying too much paint at the starting and stopping points.

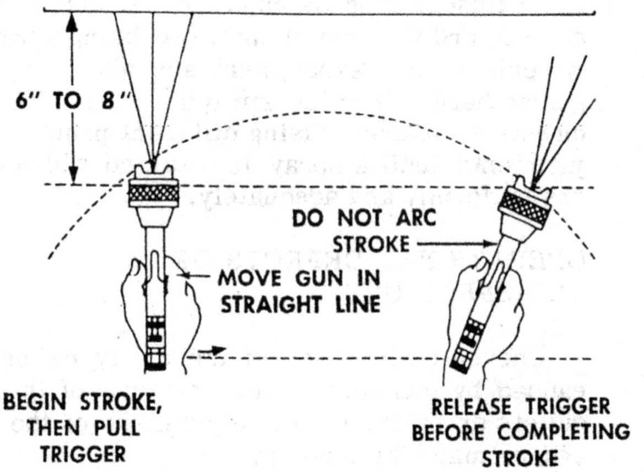

80.243

Figure 17.—Correct method of stroking with a spray gun.

Figure 18- shows right and wrong methods of spraying an outside corner. If you use the wrong method shown, a good deal of paint will be wasted into the air.

IX. AVERAGE COVERAGE OF PAINT

The area a gallon of paint will cover varies considerably with the nature of the surface, the character of the paint, and the method of application. Table 4 is intended only to give you a rough estimate of the average coverage per gallon for brush painting.

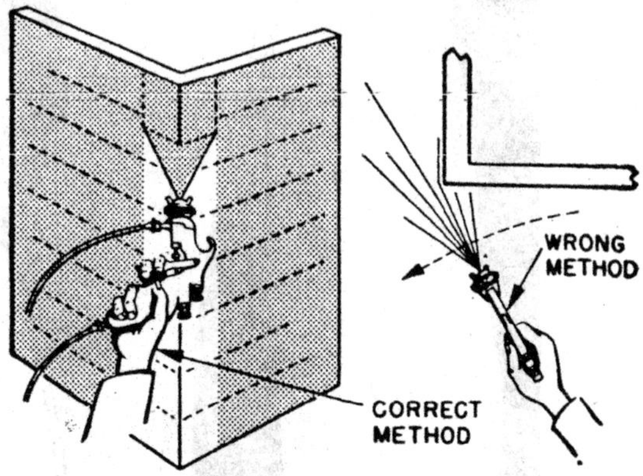

Figure 18 Right and wrong methods of spraying an outside corner.

To plan the work of your crew more competently and, particularly, to make the most effective use of your spray painting teams, you should develop the ability to estimate the number of man-hours and amount of paint required to do the ordinary painting jobs.

It is difficult to list more than a few guidelines for estimating and impossible to lay down any hard and fast rules because of the many variables involved (type of structures, skill of the team, type of paint to be used, and so on).

Keep a set of notes on the jobs that your men do. They will help you with future estimations. Note such things as the number of square feet a gallon of different types of paint will cover when applied by different methods, how much time is required to ready the equipment for spraying, how many square feet of surface a team can paint in an hour, the number of gallons of paint required for each structure and so on.

For example:

Paint coverage per gallon:

1. Enamel—400 ft^2 (square feet) (by brush)
2. Enamel—425 ft^2 (spray)
3. Flat—400 ft^2 (by brush)
4. Flat—430 ft^2 (spray)

Table 4.—Average Paint Coverage

Type of Surface		Area in Square Feet per Gallon		
		Primer or First Coat	Second Coat	Third Coat
Exterior				
Wood Siding and Trim				
	Flat Oil	300	350	400
Shingle	Stain	80	125	
Concrete Masonry Unit				
	Cement Base	100	150	200
	Latex	150	200	250
Interior				
Plaster	Flat Oil	300	350	400
	Gloss Oil	300	350	400
	Latex	300	350	400
Concrete Masonry Unit				
	Cement Base	100	150	200
	Latex	150	200	250

Team #1 (experience_______________)
Readying equipment__hr. (__helpers)
Average__ft^2 per hr.

COLORS

The mixing and use of colored paints to produce a desired color scheme or harmony is technical in nature and difficult to do well. Much could be written about colors, but unless a man works with them and gains experience in this field, he will never fully understand how to use the different colored paints.

MIXING

As you know, there are three primary colors—red, blue, and yellow. These are the only true colors and are the basis for all subsequent shades, tints, and hues that are derived by mixing any combination of these colors in various proportions. Figure 19 illustrates a color triangle with one primary color at each of its points. The lettering in the triangle indicates

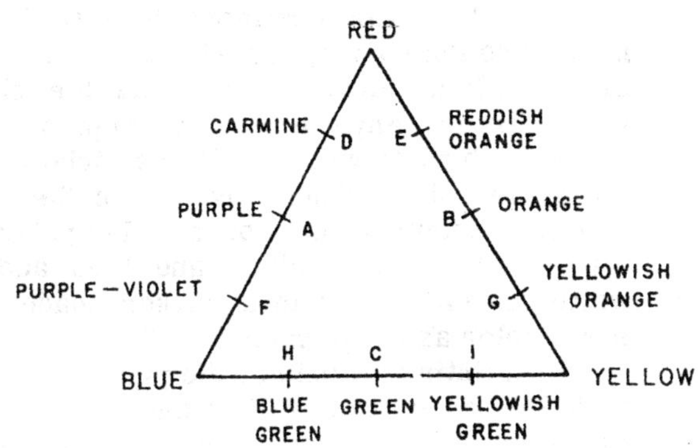

Figure 19.—A color triangle.

the hues that will result when these colors are mixed together as follows:

Equal proportions of red and blue produce a purple.

Equal proportions of red and yellow produce an orange.

Equal parts of blue and yellow produce a green.

Three parts of red to one part of blue will make a carmine.

Three parts of red to one of yellow will result in a reddish-orange.

Three parts of blue to one of red will produce a purple-violet.

Three parts of yellow to one of red will produce a yellowish-orange.

Three parts of blue to one of yellow will result in a blue-green.

Three parts of yellow to one of blue will produce a yellowish-green.

Many other hues may be produced by varying the proportions of the primary colors.

Hues are known as chromatic colors, while black, white, and gray series are known as the achromatic (neutral) colors. The gray series can be produced by mixing black and white in different proportions.

TINTING

Tinting is the process of mixing a colored pigment with a vehicle such as linseed oil, turpentine, or varnish, then blending this combination with a base paint to produce the desired shade or tint.

Some of the most commonly used tinting pigments are lampblack, raw sienna, Venetian red or vermilion, ultra-marine, and the chrome yellow group. Any one of these pigments when mixed with white will result in a lighter tint of the same color. When any one of these pigments (excepting, of course, lampblack) is mixed with a little black and then added to white, it will result in a darker shade of the same color as the pigment.

When tinting oil paints, the pigment is mixed with linseed oil to thin it to the consistency of canned milk. If an enamel paint is to be tinted, the pigment is mixed with varnish or linseed oil to the same consistency as is used for oil paints. After the tint is thoroughly mixed, it should be strained to remove any remaining lumps of pigments. If this isn't done, dark streaks will appear in the tinted paint when it is applied to a surface.

The tint should be added a little at a time to the material being tinted and the mixture spread on a test board so that its color may be compared to the color of the surface being matched. There are differences in the tinting strength of the various brands of tinting colors, and between the colors of one brand. For example: Prussian blue is stronger in tinting strength than other blue tint colors, chrome yellow is stronger than raw sienna or yellow ochre, and vermilion is much stronger than Venetian red. This refers only to their ability to tint or color white paint. However, this does not mean that a satisfactory tinting job cannot be done by using any of the tinting pigments, provided the proper amount of tinting material is blended with the base.

The following are good points to remember when tinting.

1. Colors always appear darker in the mixing can than on a surface.
2. Artificial light causes colors to appear darker than they are in daylight.
3. All colors dry to a lighter shade than they appear before being applied to a surface. Paints should be mixed to a slightly darker shade than the color being matched.
4. The eyes easily become saturated with a color and the colors may seem to fade or change. To assure accuracy when you are mixing colors, it is a good policy to look away from the work for a few minutes and rest your eyes.

X. DETERIORATION OF PAINT

Paints are not indestructible. Even properly selected protective coatings properly applied on well prepared surfaces will gradually deteriorate and eventually fail. The rate of deterioration under such conditions, however, is slower than when improper painting operations are carried out. Inspectors and personnel responsible for maintenance painting must be familiar with the signs of various stages of deterioration in order to establish an effective and efficient system of inspection and programmed painting. Repainting at the proper time avoids the problems resulting from painting either too soon or too late. Painting scheduled before it is necessary is uneconomical and eventually results in a heavy film buildup leading to abnormal deterioration of the paint system. Painting scheduled too late results in costly surface preparation and may be responsible for damage to the structure, which then may require expensive repairs.

A paint which reaches the end of useful life PREMATURELY is said to have FAILED. The following sections describe the more common types of paint failures, the reasons for such failures, and methods of prevention and/or cure.

CHALKING

Chalking is the result of weathering of the paint at the surface of the coating. The vehicle is broken down by sunlight and other destructive influences, leaving behind loose powdery pigment which can easily be rubbed off with the finger. (See fig. 20.) Chalking takes place more rapidly with softer paints such as those based on linseed oil. Chalking is most rapid in areas exposed to large amounts of sunshine. For example, in the northern hemisphere, chalking will be most rapid on the south side of a building. On the other hand, little chalking will take place in areas protected from sunshine and rain such as under eaves or overhangs. Controlled chalking can be an asset, especially in white paints, since it is a self cleaning process and helps to keep the surface clean and white. Furthermore, by gradually wearing away, it reduces the thickness of the coating, thus allowing continuous repainting without making the coating too thick for satisfactory service. Chalked paints are also generally easier to repaint since the underlying paint is in good condition, and, generally, little surface preparation is required. This is not the case when water-thinned paints are to be applied. Their adhesion to chalky surfaces is poor.

ALLIGATORING

Alligatoring describes a pattern in a coating which looks like the hide of an alligator. It is caused by uneven expansion and contraction of a relatively undercoat. (See fig. 21.) Alligatoring can be caused by:

1. applying an enamel over an oil primer
2. painting over bituminous paint, asphalt, pitch or shellac
3. painting over grease or wax.

PEELING

PEELING (fig. 22) results from inadequate bonding of the top coat with the undercoat or the underlying surface. It is nearly always caused by inadequate surface preparation. A top coat will peel if it is applied to a wet surface, a dirty surface, an oily or waxy surface, or a glossy surface. All glossy surfaces must be sanded before painting.

BLISTERING

BLISTERING (fig. 23) is caused by the development of gas or liquid pressure under the paint. The root cause of most blistering, other than that caused by excessive heat, is inadequate ventilation plus some structural defect that allows moisture to accumulate under the paint. Before repainting, the cause of the blistering must be determined and corrected. Blisters should be scraped off, the paint edges around them should be feathered off with sandpaper, and the bare places primed before the blistered area is repainted.

CHECKING AND CRACKING

Checking and cracking describe breaks in the paint film which are formed as the paint becomes hard and brittle. Temperature changes cause the substrate and overlying paint to expand and contract. As the paint becomes hard, it gradually loses its ability to expand without breaking to some extent. Checking is described as tiny breaks which take place only in the

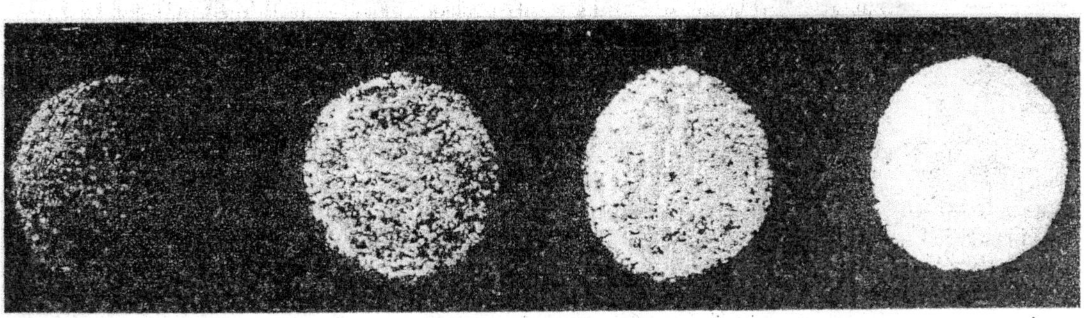

Figure 20.—Degrees of chalk.

Figure 21.—Alligatoring.

Figure 22.—Peeling.

upper coat or coats of the paint film without penetrating to the substrate. The pattern is usually similar to a crowsfoot. (See fig. 24.) Cracking describes larger and longer breaks which extend through to the substrate. (See fig. 25.) Both are a result of stresses in the paint film which exceed the strength of the

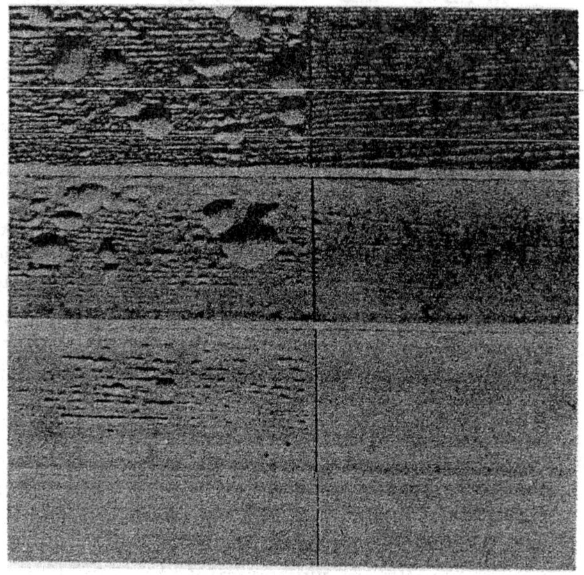

Figure 23.—Blistering.

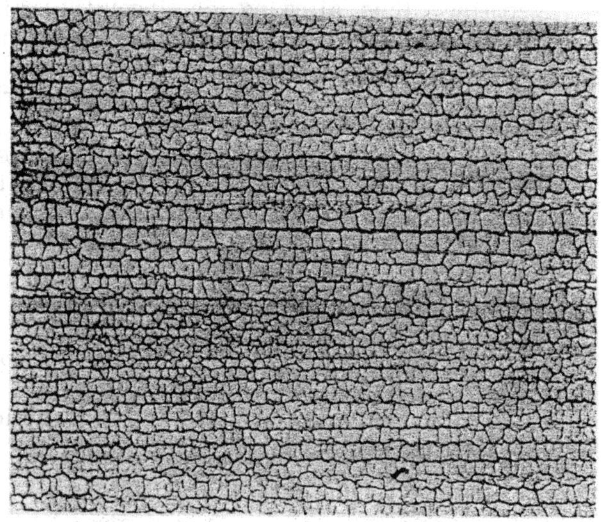

Figure 24.—Severe checking.

coating. Whereas checking arises from stresses within the paint film, cracking is caused by stresses between the film and the substrate. Cracking will generally take place to a greater extent on wood than on other substrates because of its grain. When wood expands, it expands much more across the grain than along the grain. Therefore, the stress in the coating is greatest across the grain causing cracks to

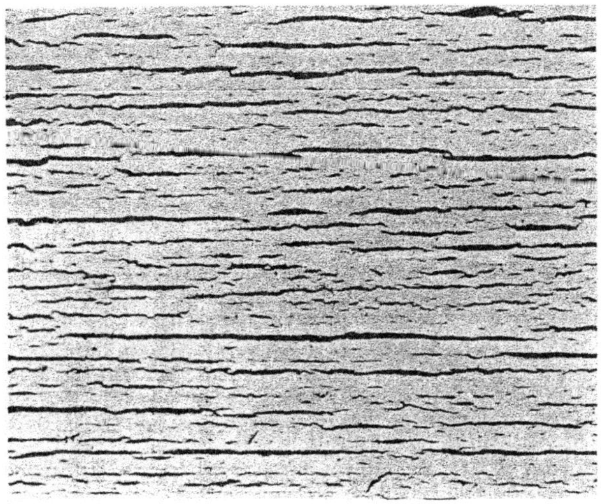

Figure 25.—Severe cracking.

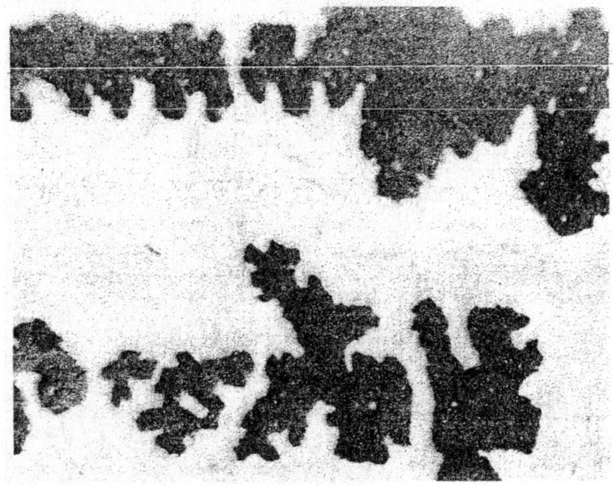

Figure 26.—Crawling.

form parallel to the grain of the wood. Checking and cracking are aggravated by excessively thick coatings because of their reduced elasticity.

CRAWLING

Crawling occurs when the new coating fails to wet and form a continuous film over the preceding coat. Examples are applying latex paints over high gloss enamel or applying paints on concrete or masonry treated with a silicone water repellant. (See fig. 26.)

INADEQUATE GLOSS

Sometimes a gloss paint fails to attain the normal amount of gloss. This may be caused by (1) inadequate surface preparation, (2) application over an undercoat which is not thoroughly dry, (3) application in cold or foggy weather.

PROLONGED TACKINESS

A coat of paint is dry when it ceases to be "tacky" to the touch, and prolonged tackiness indicates excessively slow drying. This may be caused by (a) insufficient drier in the paint, (b) a low-quality vehicle in the paint, (c) applying the paint too thickly, (d) painting over an undercoat which is not thoroughly dry, (3) painting over a waxy, oily, or greasy surface, and (f) painting in damp, wet, or foggy weather.

WRINKLING

When paint is applied too thickly, especially in cold weather, the surface of the coat dries to a skin over a layer of undried paint underneath. This usually causes WRINKLING like that shown in figure 27. To avoid wrinkling when you are brush-painting or roller-painting, be sure to brush or roll each coat of paint out as thin as possible. To avoid wrinkling when spray painting, be careful to keep the gun in constant motion over the surface whenever you have the trigger down.

XI. PAINTING SAFETY

Every painting assignment exposes Builders to conditions and situations that represent actual or potential danger to themselves and to others in the area. The frequent necessity to use toxic and flammable materials, pressurized equipment, ladders, scaffolding and rigging always presents a potential hazard. Hazards may also be inherent in the very nature of the environment, or caused through ignorance or carelessness of the operator. It is, therefore, extremely important to be aware of all potential hazards, since continuous and automatic precautionary measures will minimize the problem and improve both efficiency and morale of the painting crew.

Painting accident hazards may be broadly divided into three major types as follows:

Figure 27.—Wrinkling.

1. Hazards involved in the use of scaffolds, ladders, and rigging equipment.
2. Fire hazards from flammable materials in paints.
3. Health hazards from toxic (poisonous) materials in paints.

EQUIPMENT HAZARDS

Accidents during painting operations are caused by unsafe working equipment, unsafe working conditions and careless personnel.

Nothing should be taken for granted. Proper use of equipment must be taught to all personnel by qualified Builders. Refresher courses on the use of all equipment must be regularly scheduled.

The following basic procedures in setting up and use of equipment are imperative to assure safety standards and maximum protection of all personnel.

Ladders

1. Store wood ladders in a warm, dry area protected from the weather and ground.
2. Protect wood ladders with clear coatings only, so that cracks, splinters or other defects will be readily visible.
3. Inspect all ladders frequently for loose or bent parts, cracks, breaks or splinters.
4. All straight and extension ladders must have safety shoes. These should be of insulating material for metal ladders. (See fig. 28.)

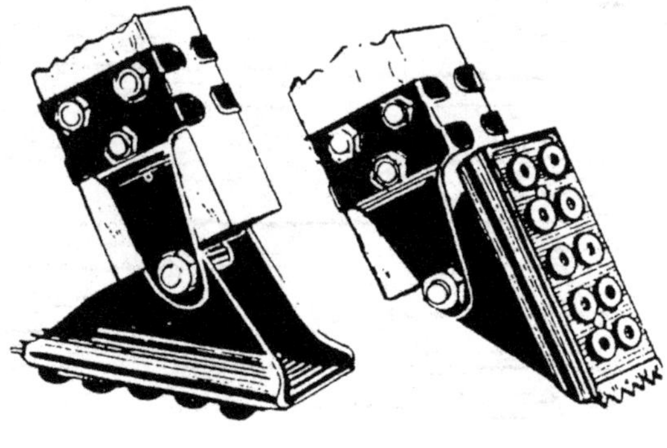

Figure 28.—Ladder safety shoes.

5. Do not use portable ladders greater in length than can be readily carried and placed by two men. Never splice ladders to form a longer ladder.
6. Pre-test all ladders and scaffolding before use by placing horizontally with blocks under ends and "bouncing" in the center or walking along ladder or scaffold.
7. Extension ladders should have a minimum overlap of 15% for each section. (See fig. 18-29.)
8. Do not use stepladders over 12 feet high. Never use one as a straight ladder. Never stand on the top platform.
9. Place ladders so that the horizontal distance from the top support to foot is at least 1/4 of the working length. Be sure that the ladder is securely in place. Rope off all doorways in front of the ladder and place warning signs.
10. Use hand lines to raise or lower tools and materials. Do not overreach when working on ladders. Move the ladder instead.
11. Never use metal ladders in areas where contact with electric power lines is possible.

Scaffolding

1. Inspect all parts before use. Reject metal parts damaged by corrosion and wood parts with defects such as checks, splits, unsound knots and decay.
2. Provide adequate sills or under-pinnings when erecting on filled or soft ground. Be sure that scaffolds are plumb and level. Compensate

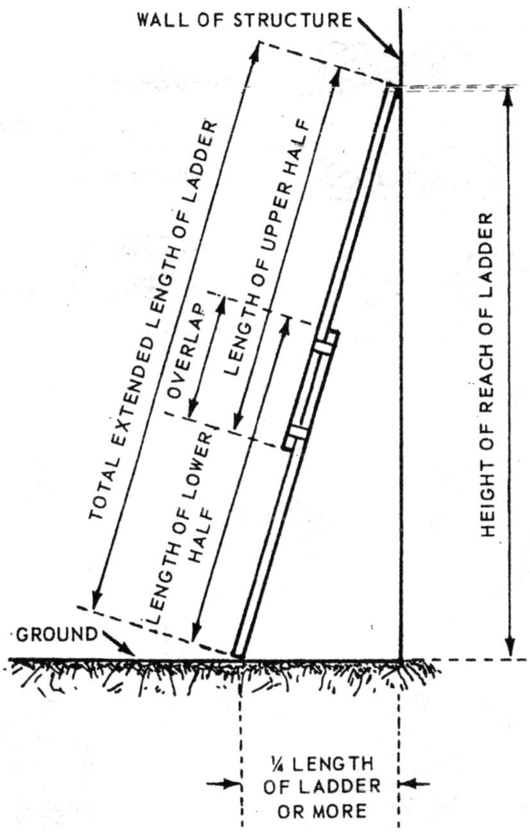

Figure 29.—Ladder stability.

for unevenness of the ground by blocking or using adjusting screws.

3. Anchor scaffolds to the wall about every 28 feet in length and 18 feet in height. Do not force braces to fit. Use horizontal diagonal bracing at bottom and at every 30 feet in height.

4. Lumber should be straight grained. All nails should be driven full length and not subject to direct pull.

5. Provide guard railings regardless of height, on the full length of the scaffold and also on the ends.

6. Erect scaffolding so that ladders are lined up from top to bottom. Always use ladders when climbing scaffolding.

7. Tubular pole scaffolds should be made of two inch O.D. galvanized steel tubing or other corrosion resistant metal of equal strength. They should be erected or dismantled by experienced Builders only.

8. Planking should have at least a two foot overlap. Secure well to wood scaffolding. Platforms should be made of planking of uniform thickness laid close together. They must overlap and be fastened at supports. They must not extend over the edge without being properly supported. An unsupported plank is a deadly trap. Do not use planking for other purposes; paint them only at the ends to identify them. Nominal sizes of planking should be determined from table 5. Values are given in pounds for loads at center and allow for weight of planking.

9. Test scaffolds and extensive planking (extended to working length) by raising them one foot off the ground and loading them with weights at least 4 times the anticipated working load.

Table 5.—Safe Center Loads for Scaffold Plank.

Span feet	*2 x 8**	*2 x 10**	*2 x 12**	*3 x 8**	*3 x 10**	*3 x 12**
6	200	255	310	525	665	805
8	150	190	230	390	500	605
10	120	155	185	315	400	485
12	100	130	155	265	335	405
14	—	110	135	225	285	346
16	—	—	115	195	250	305

Above values are for planks supported at the ends, wide side of plank face up, and with loads concentrated at the center of the span. For loads uniformly distributed on the wide surface throughout the length, the safe loads may be twice those given in the table. Loads given are net and do not include the weight of the plank. If select structural coast region Douglas fir, merchantable structural longleaf southern pine, or dense structural square edge sound southern pine are used, above loads may be increased 25 percent.

* Dressed sizes of planks, reading left to right, are: 1⅝ x 7½, 1⅝ x 9½, 1⅝ x 11½, 2⅝ x 7½, 2⅝ x 9½, 2⅝ x 11½, respectively.

Rolling Towers

1. Inspect all tower parts before use. Do not use parts which are damaged by corrosion, deterioration, or misuse.

2. Guy or tie off towers with heights more than three times the minimum base dimension, and fix towers at every 18 feet of elevation. Maintain stability of towers over 25 feet high with out-riggers or handling lines. Use horizontal diagonal bracing at bottom and at every height section.

3. Provide unit lock arms on all towers. Do not use casters less than 6 inches in diameter. Do not extend adjusting screws more than 12 inches.

4. Do not ride towers. Look where you are going when moving them. Do not attempt to move a tower without sufficient help. Apply all caster brakes when tower is stationary.

Swinging Scaffolds,
Swing Stages, Bosun Chairs

1. Always read instructions on the proper use and maintenance of the equipment. Follow prescribed load capacities.

2. Stages should be at least 27 inches wide and supplied with guard rails (not rope).

3. Only experienced Builders are to erect or operate stages. Check ropes and blocks before use by suspending stages one foot off the ground and loading at least 4 times the anticipated work load. Before locating on the job site, check for nearby electric power lines.

4. Power stages should have free fall safety devices with hand controls in case of power failure.

Ropes and Cables

1. Store ropes and cables coiled in dry empty drums.

2. Use wire rope at least 3/4 inch diameter for platform slings; use manila rope at least 5/8 inch diameter in bosun chairs and life lines. Use proper clamps with wire rope, and proper knots and hitches when handling materials with manila rope. (See fig. 30.)

3. Inspect ropes frequently. Discard if exposed to acid or excessive heat. Check for dry rot, brittleness or excessive wear. Never use frozen rope.

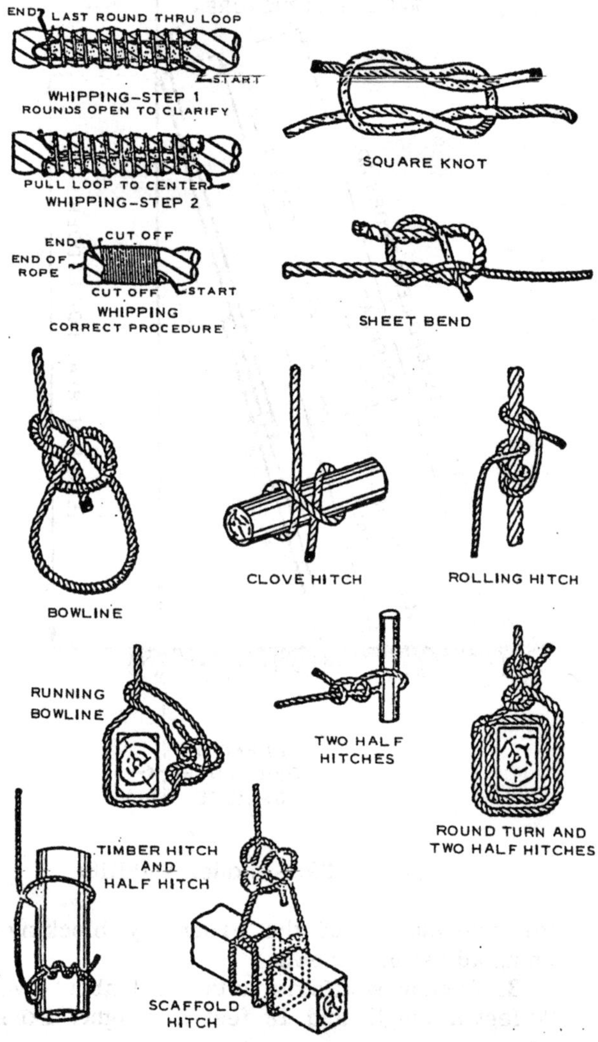

Figure 30.—Knots and hitches used in painting operations.

4. Inspect all wire ropes and cables frequently in accordance with current service safety criteria.

5. Do not attempt to salvage rope or cable by splicing.

Pressurized Equipment

These rules apply to all types of equipment used both for spraying and blasting.

1. Use only approved equipment. Use remote control deadman valves on high pressure equipment (60 lbs. or higher). These should be activated by the same air used for blasting or spraying.

2. Conduct a hydrostatic test at least once, preferably twice a year. Test safety relief valves daily.

3. Use conductive hose. Ground nozzles, tanks and pressure equipment when in use, also object being sprayed.

4. Store hose in dry areas. When in use, avoid sharp bends, especially when curved around an object. Secure high pressure hose no more than 10 ft. from operator.

5. Never point gun or nozzle at anyone or any part of the body. When handling or carrying, hold by the grip and remove the fingers from the trigger.

6. Release all pressure before disconnecting any part of the equipment.

FIRE HAZARDS

Certain general rules regarding fire and explosion hazards apply to all situations. All paint materials should have complete label instructions which stipulate the potential fire hazards and precautions to be taken. Painters must be continuously advised and reminded of the fire hazards that exist under the particular conditions of each job, so that they will be aware of the dangers involved and assure that the necessary precautions are taken and maintained. Fire fighting equipment, of the proper type, must always be readily available in the paint shop, spray room and work areas where a potential fire hazard exists. Electric wiring and equipment installed or used in the paint shop, including storage room and spray room, must conform to the applicable requirements of the National Electrical Code for Hazardous Areas. The following precautions against fire must be carefully observed by all paint-handling personnel:

1. Prohibit smoking anywhere that paint is either stored, prepared for use or applied.

2. Provide for adequate ventilation in all of these areas.

3. Perform recurrent spray operations on portable items, e.g., signs, in an approved spray booth equipped with adequate ventilation, a water wash system of fume removal and explosion proof electrical equipment.

4. Wet down spray booth surfaces before cleaning them.

5. Use rubber feet on metal ladders, and be certain that personnel working in hazardous areas use rubber soled shoes.

6. Use non-sparking scrapers and brushes to clean metal surfaces where fire hazards are present.

7. Wet down paint sweepings, rags and waste with water, and store in closed metal containers until disposed of in an approved manner. Do not burn in heaters or furnaces. (See fig. 31.)

Figure 31.—Keep combustible materials in metal waste cans tightly covered.

8. Extinguish all pilot lights on water heaters, furnaces and other open flame equipment on all floors of the structure being painted. Be sure to turn the gas valve off.

9. When painting in confined areas near machinery or electrical equipment, open all switches and tag them to prevent their being turned on inadvertently.

10. Be sure that all mixers, pumps, motors, and lights used in the paint shop, spray room or on the job are explosion proof and electrically grounded.

11. Use pails of sand (never sawdust) near dispensing pumps and spigots to absorb any spillage or overflow.

12. During painting operations keep fire extinguishers nearby. Be sure that they are of the proper type. (See table 6.)

13. Check ventilation and temperature regularly when working in confined areas.

14. Consult with the CEs before painting in areas where high voltage lines and equipment are located.

15. Keep all work areas clear of obstructions.

16. Clean up before, during and after painting operations. Dispose of sweepings and waste daily.

HEALTH HAZARDS

A variety of ingredients used in the manufacture of paint materials are injurious to the human body in varying degrees. While the body can withstand nominal quantities of most of these poisons for relatively short periods of time, continuous or over exposure to them may have harmful effects. Furthermore, continued exposure to some may cause the body to become sensitized so that subsequent contact, even in small amounts, may cause an aggravated reaction. To this extent, these materials are a very definite threat to the normally healthy individual and a serious danger to persons with chronic illnesses or disorders. These materials are divided into two major groups, i.e., toxic materials and skin irritating materials.

Nevertheless, health hazards can easily be avoided by a common sense approach of avoiding unnecessary contact with hazardous materials

Table 6.—Use the Proper Fire Extinguisher

	Three Classes of Fires		
Choose from these 5 basic types of extinguishers	**CLASS A FIRES** Paper, wood, cloth, excelsior, rubbish, etc., where quenching and cooling effect of water is required.	**CLASS B FIRES** Burning liquids (gasoline oil, paints, cooking fats, etc.) where smothering action is required.	**CLASS C FIRES** Fires in live electrical equipment (motors, switches, appliances, etc.) where a non-conducting extinguishing agent is required.
CARBON DIOXIDE	Small surface fires only.	**YES Excellent** Carbon dioxide leaves no residue, does not affect equipment or foodstuffs.	**YES Excellent** Carbon doxide is a non-conductor, leaves no residue, will not damage equipment.
DRY CHEMICAL	Small surface fires only.	**YES Excellent** Chemical absorbs heat and releases smothering gas on fire; chemical shields operator from heat.	**YES Excellent** Chemical is a non-conductor; fog of dry chemical shields operator from heat.
WATER	**YES Excellent** Water saturates material and prevents rekindling.	**NO** Water will spread fire, not put it out.	**NO** Water, a conductor, should not be used on live electrical equipment.
FOAM	**YES Excellent** Foam has both smothering and wetting action.	**YES Excellent** Smothering blanket does not dissipate, floats on top of most spilled liquids.	**NO** Foam is a conductor and should never be used on live electrical equipment.
VAPORIZING LIQUID	Small surface fires only.	**YES** Releases heavy smothering gas on fire.	**YES** Liquid is a non-conductor and will not damage equipment.

and by strict adherence to established safety measures.

The following rules should always be strictly observed:

1. Toxic or dermatitic materials must be properly identified and kept tightly sealed when not in use.

2. Designate a competent person to check the operation of paint spray booths. Check at regular intervals to ensure that the equipment is in a safe and proper operating condition.

3. Be sure that ventilation is adequate in all painting areas. Provide artificial ventilation where natural ventilation is inadequate. Use supplied air respirators, if necessary.

4. Spray all portable items within exhaust ventilated booths especially designed for that purpose.

5. Wear goggles and the proper type of respirator when spraying, blast cleaning or performing any operation where any abnormal amount of vapor, mist or dust is formed.

Table 7.—Recommended Preservatives and Retentions for Ties, Lumber, Piles, Poles, and Posts

Product	Minimum net retention of—(Pounds per cubic foot)				
	Coal-tar creosote	Creosote-coal tar solution	Creosote-petroleum solution	Pentachlorophenol, 5 percent in petroleum	Copper naphthenate (0.75 percent copper metal) in petroleum
Ties (crossties, switch ties, and bridge ties)	8	8	9		
Lumber, and structural timbers:					
For use in coastal waters:					
Douglas fir (coast type) lumber and timbers.	14	14			
Southern yellow pine lumber and timbers	20	20			
For use in fresh water, in contact with ground or for important structural members not in contact with ground or water.	10	10	12	10	10
For other use not in contact with ground or water.	6	6	7	6	6
Piles:					
For use in coastal waters:					
Douglas fir (coast type). . . .	17	17			
Southern yellow pine	20	20			
For land or fresh water use . .	12	12	14	12	
Poles (utility and building)	8, 10	. . . .	. . . .	8, 10	8, 10
Posts (round, fence)	6	6	7	6	6

6. When handling dermatitic materials, use protective creams or preferably gloves, and wear appropriate clothing. Change and clean work clothing regularly.

7. Avoid touching any part of the body, especially the face, when handling dermatitic materials. Wash hands and face thoroughly before eating and at the end of the day.

XII. WOOD PRESERVATIVE

Damage to building and other structures by termites, wood bores, and fungi, is needless waste. Defects in wood have been caused by improper care after preservation treatment. All surfaces of treated wood that are cut or drilled to expose the untreated interior must be treated with the proper application of wood preservative.

The capacity of any wood to resist dry rot, termites, and decay, can be greatly increased by impregnating the wood with a general-purpose wood preservative or fungicide. Prescribed preservatives are listed in tables 7 and 8.

Different woods have different capacities for absorbing preservative or other liquids, and in any given wood the sapwood is much more absorbent than the heartwood. Hardwoods are, in general, less absorbent than softwoods. Naturally, the extent to which the preservative affords protection increases directly with the distance to which it penetrates below the surface of the wood. The best penetration is obtained by a pressure process which requires equipment you will not have available. Nonpressure methods of applying preservatives are by dipping and by ordinary surface application with a brush or spray gun.

Figure 32 shows how you can improvise long tanks for the dipping process. Absorption is rapid at first, much slower later. A rule of thumb is to the effect that in 3 minutes wood will have absorbed one-half the total amount of preservative it will absorb in 2 hours. However, the extent of the penetration obtained will

Table 8.–Minimum Retentions of Water-Borne Preservatives

Preservative	Minimum retentions for uses	
	Not in contact with ground or in water	Involving occasional exposure to rainwater or continually to ground in areas of low rainfall
	Pounds per cubic foot	Pounds per cubic foot
Acid copper chromate	0.5	1.00
Ammoniacal copper arsenite	.3	.5
Chromated copper arsenate	.35	.75
Chromated zinc arsenate (including copperized form)	.5	1.00
Chromated zinc chloride	.75	1.00
Copperized chromated zinc chloride	.75	1.00
Fluor-chrome-arsenate-phenol	.35	.50

depend upon the type of wood, its moisture content, and the length of time it remains immersed.

Surface application by brush or spray is, from the standpoint of a desire for maximum penetration, the least satisfactory method of treating wood. However, it is more or less unavoidable in the case of any wood which is already installed, as well as for treated wood which has been cut or drilled to expose the untreated interior.

Pentachruphenol and creosote coal tar are likely to be the only field-mixed preservative used by the Builder. The type of treatment or preservative depends on the severity of exposure and the desired life of the end product. Types and uses of wood preservatives are shown in tables 7 and 8. However, the Builder should be familiar with the following safety precautions:

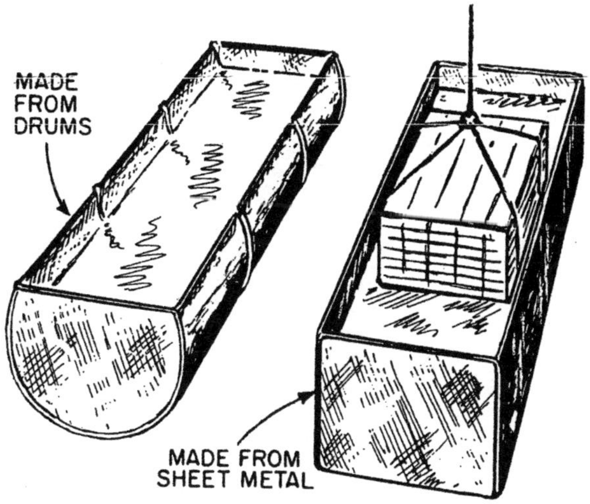

Figure 32.—Improvised tanks for dip-treating lumber.

1. Avoid undue skin contact.
2. Avoid touching the face or rubbing the eyes when handling pretreated material.
3. Avoid inhalation of toxic material.
4. The application of preservative is very hazardous; apply only in a properly ventilated space and use approved respirators.
5. Wash with soap and water after contact.

GLOSSARY OF PAINTING TERMS

TABLE OF CONTENTS

GLOSSARY OF PAINTING TERMS

Painting operations employ terms that are peculiar to this field and, as such, may require some explanation or definition. This glossary is designed to provide the reader with some basic understanding of terms commonly used in painting and thus eliminate possible misunderstandings resulting from conflicting interpretations of terms and improve communication between all persons involved in the painting operation.

A

abrasive—the agent used for abrasive blast cleaning; for example, sand, grit, steel shot, etc.

absorption—process of soaking up, or assimilation of one substance by another.

accelerator—catalyst; a material which accelerates the hardening of certain coatings.

acetone—a fast evaporating, highly flammable solvent.

acoustic paint—paint which absorbs or deadens sound.

acrylic resin—a clear resin derived from polymerized esters of acrylic acid and methacrylic acid.

activator—catalyst or curing agent; accelerator.

adhesion—bonding strength, the attraction of a coating to the surface to which it is applied.

absorption—process of attraction to a surface; attachment; the retention of foreign molecules on the surface of a substance.

adulteration—the addition of unwanted materials.

agglomeration—formation of masses or aggregates of pigments; not dispersed.

air adjusting valve—spray gun valve controlling input air.

air bubble—bubble in paint film caused by entrapped air.

air cap (or air nozzle)—perforated housing for atomizing air at head of spray gun.

air drying—drying by oxidation or evaporation by simple exposure to air.

air entrapment—inclusion of air bubbles in paint film.

air hose—hose of air supply quality, usually red.

air jet (blast cleaning)—type of blast cleaning gun in which the abrasive is conveyed to the gun by partial vacuum.

airless spraying—spraying using hydraulic pressure to atomize the paint.

air manifold—common air supply chamber for several lines.

air transformer—device for controlled reduction in air pressure.

air valve—control valve in air line system.

air volume—quantity of air in cubic feet (usually per minute) at atmospheric pressure.

alcohol—a flammable solvent; alcohols commonly used in painting are ethyl alcohol (ethanol) and methyl alcohol (methanol, wood alcohol).

aliphatic hydrocarbons—flammable solvents of low solvent power, usually derived from petroleum.

alkali—caustic, such as sodium hydroxide, lye, etc.

alkyd resins—resins prepared from polyhydric alcohols and polybasic acids.

alligatoring—surface imperfections of paint having the appearance of alligator hide.

ambient temperature—room temperature or temperature of surroundings.

American gallon—231 cubic inches.

amides—curing agent combined with epoxy resins.

amines—curing agent combined with epoxy resins.

anchor pattern—profile of a surface, usually attained by blasting.

angle blasting—blast cleaning at angles less than 90 degrees.

angle of degree (airless spray cap)—orifice angle; controls width of spray, pattern angle.

anhydrous—dry, free of water in any form.

applicator—one who applies; tool for applying.

arcing—swinging spray gun away from the work.

aromatic hydrocarbons—strong solvents such as benzene, toluene, xylene.

asphalt—residue from petroleum refining; also a natural complex hydrocarbon.

atomize—break stream into small particles.

B

baking finish—product requiring heat cure.

banding—identifying with strips of tape.

barrier coating—shielding or blocking coating or film.

binder—resin; drying oil; latex emulsion; film former; vehicle.

bituminous coating—coal tar or asphalt based coating.

blast angle—angle of nozzle with reference to surface; also angle of particle propelled from rotating blast cleaning wheel with reference to surface.

blast cleaning—cleaning with propelled abrasives.

bleaching—removing color.

bleeder gun—a spray gun with no air valve; trigger controls fluid flow only.

bleeding—penetration of color from the underlying surface.

blisters—bubbles in dry or partially-dry paint film.

blooming—whitening; moisture blush; blushing.

blow-back (spray term)—rebound of atomized sprayed material.

blushing—whitening and loss of gloss due to moisture or improper solvent balance.

body—viscosity; middle or under coat; to thicken.

boilers (solvent)—solvents of particular evaporation rate.

bonding—adhesion

bounce-back—spray rebound similar to blow-back.

boxing—mixing by pouring back and forth from one container to another.

bridging—forming a skin over a depression.

bright blast—white blast; See 4.4.2.4.

brittleness—degree of resistance to cracking, breaking or bending.

broadcast—to sprinkle solid particles on a surface.

bronze tools—non-sparking tools; used when fire hazards are particularly acute.

bronzing—formation of metallic sheen on a paint film.

brushability—ability to be brushed.

brush-off blast—lowest blast cleaning standard; see 4.4.2.4.

bubbling—a term used to describe the appearance of blisters on the surface while a coating is being applied.

C

caking—hard settling of pigment from paint.

camouflage—the art or system for deception or concealment.

catalyst—accelerator; curing agent; promoter.

cat-eye—hole or holiday shaped like a cat's eye; cratering.

chalking—powdering of surface.

checking—formation of slight breaks in the film that do not penetrate to the underlying surface.

chipping—(1) cleaning steel using special hammers.
—(2) type of paint failure.

chlorinated rubber—a particular film former used as a binder, made by chlorinating natural rubber.

cleaner—(1) detergent, alkali, acid or other cleaning material; usually water or steam borne.
(2) solvent for cleaning paint equipment.

coal tar pitch—black residue remaining after coal tar is distilled.

coal tar-epoxy paint—paint in which binder or vehicle is a combination of coal tar with epoxy resin.

coatings—surface coverings; paints; barriers.

coat of paint—layer of dry paint resulting from a single wet application.

cobwebbing—a spider web effect caused by premature drying.

cohesion—property of holding self together.

cold-checking—checking caused by low temperatures.

cold-cracking—cracking occuring at low temperatures.

color-fast—non-fading.

color retention—ability to retain original color.

commercial blast—see 4.4.2.4.

compatibility—ability to mix with or adhere properly to other components or substances.

composition—analysis; make-up.

conditioner—see surface conditioner.

continuity—degree of being intact or pore free.

copolymer—large molecule resulting from simultaneous polymerization of different monomers.

copper sulfate test (for mill scale)—copper color indicates absence of mill scale when steel is swabbed with 5 to 10 per cent solution.

corrosion—oxidation; deterioration due to interaction with environment.

cracking—splitting, disintegration of paint by breaks through film to substrate.

cratering—formation of holes or deep depressions in paint film.

crawling—shrinking of paint to form uneven surface shortly after application.

crazing—development of non-uniform surface appearance of myriad tiny scales or cracks.

creepage—see crawling.

cross-linking—a particular method by which chemicals unite to form films.

cross-spray—spraying first in one direction and then at right angles.

curing—setting up; hardening.

curing agent—hardener; promoter.

curtaining—sagging.

curtains—sags having appearance of drapes.

cycling (of pump)—interval between strokes.

D

deadman valve—shut-off valve at blast nozzle, operated by remote control.

decorative painting—painting for appearance.

degreaser—chemical solution (compound) for grease removal.

delamination—separation of layers of paint films.

density—weight per unit volume.

detergent—cleaning agent.

dew point—temperature at which moisture condenses.

diluents—see thinners.

discoloration—color change.

dispersion—suspension of one substance in another.

distensibility—ability to be stretched.

distillation—purification or separation by volatilizing and condensing.

doctor blade—knife applicator of fixed film thickness.

double regulation—regulation of both pot and gun air pressure.

drier—chemical which promotes oxidation or drying of paint.

drift (overspray)—spray loss.

drop (scaffold)—one vertical descent of the scaffold.

drop cloth—protective cover.

dry film thickness—depth of applied coating when dry, expressed in mils (1/1000 in.).

dry spray—overspray or bounce back; sand finish due to spray particle being partially dried before reaching the surface.

drying oil—an oil which hardens in air.

drying time—time interval between application and a specified condition of dryness.

dry to handle—time interval between application and ability to be picked up without damage.

dry to recoat—time interval between application and ability to receive next coat satisfactorily.

dry to touch—time interval between application and ability to be touched lightly (tack-free time).

dulling—loss of gloss or sheen.

E

edging—striping.

efflorescence—deposit of soluble white salts on surface of brick and other masonry.

eggshell—between semi-gloss and flat.

elasticity—degree of recovery from stretching.

electrostatice spray—spraying in which electric charge attracts paint to surface.

emulsion paint—water-thinned paint with an emulsified oil and/or resin or latex vehicle.

enamel—a paint which is characterized by an ability to form an especially smooth film.

epoxy resins—film formers usually made from bisphenol A and epichlorohydrin.

epoxy amine—amine cured epoxy resin.

epoxy ester—epoxy modified oil; single package epoxy.

erosion—wearing away of paint films to expose the substrate or undercoat.

estimate—compute; calculated cost of a job.

etch—surface preparation of metal by chemical means.

evaporation rate—rate at which a solvent evaporates.

evaporation rate, final—time interval for complete evaporation of all solvents.

evaporation rate, initial—time interval during which low boiling solvent evaporates completely.

explosive limits—a range of the ratio of solvent vapor to air in which the mixture will explode if ignited. Below the lower or above the higher exposive limit, the mixture is too lean or too rich to explode. The critical ratio runs from about one to twelve per cent of solvent vapor by volume at atomospheric pressure.

extender—pigment which can contribute specific properties to paint, generally low in cost.

extension gun—pole gun.

external mix—spray equipment in which fluid and air join outside of aircap.

F

fading—reduction in brightness of color.

fallout (spray)—overspray

fanning (spray gun technique)—arcing; moving the spray gun away from the work.

fan pattern—geometry or shape of spray pattern.

feather edge—tapered edge.

feathering—(1) triggering a gun at the end of each stroke; (2) tapering edge.

Federal specifications—Government specifications for products, components and/or performance.

ferrous—iron containing.

field painting—painting at the job site.

filler—extender; bulking agent; inert pigment.

film build—dry thickness characteristics per coat.

film-former—a substance which forms a skin or membrane when dried from a liquid state.

film integrity—degree of continuity of film.

film thickness—depth of applied coating, expressed in mils (1/1000 in.).

film thickness gauge—device for measuring film thickness; both wet and dry gauges are available.

filter—strainer; purifier.

fineness of grind—measure of particle size or roughness of liquid paint; degree of dispersion of pigment in the binder.

fingers (airless spray)—broken airless spray pattern.

fire-retardant paint—a paint which will delay flaming or over-heating of substrate.

fish eye—see cratering.

flaking—disintegration in small pieces or flakes; see scaling.

flammability—measure of ease of catching fire; ability to burn.

flame cleaning—method of surface preparation of steel using flame.

flash point—the lowest temperature at which a given flammable material will flash if a flame or spark is present.

flatting—loss of gloss in coating film.

flexibility—ability to be bent without damage.

floating—separation of pigment colors on surface.

flooding—see floating.

flow—a measure of self leveling.

fluid adjusting screw—a screw on a spray gun which controls the amount of fluid entering the gun.

fluid flow—a measure of flow through a gun with atomizing air shut off.

fluid hose—specially designed hose for paint materials; usually black.

fluid nozzle—fluid tip with orifice; in a broader sense it means needle and tip combination.

fluid tip—orifice in gun into which needle is seated.

foaming—frothing.

fogging—misting.

forced drying—acceleration of drying by increasing the temperature above ambient temperature using an oven, infra red lamp or other heat source.

fungicide—a substance poisonous to fungi; retards or prevents fungi growth.

furane resins—dark chemical resistant resins made from furfuryl alcohol, furfural, and phenol.

G

galvanized steel—steel plated in a molten bath of zinc.

gas checking—fine checking; wrinkling, frosting under certain drying conditions; said to be caused by rapid oxygen absorption or by impurities in the air.

gel—a jelly-like substance.

~~**gelling (gelation)**—conversion of a liquid to a gel state.~~

glazing (puttying)—setting glass.

gloss—shininess; lustre; ability to reflect in mirror direction.

gloss retention—ability to retain original gloss.

grain—surface appearance, usually of wood.

gray blast cleaning—commercial blast. See 4.4.2.4.

grind gauge—instrument for measuring degree of pigment dispersion in liquid paint. Hegman is a common proprietary instrument.

grit—an abrasive obtained from slag and various other materials.

ground wire—a wire attached to dissipate electrostatic charge in airless spraying.

guide coat—a coat similar in composition to the finish or color coat, but of a different color to help obtain complete coverage.

gun distance—space between tip of gun and work.

H

hardener—curing agent; promoter; catalyst.

hardness—the degree to which a material will withstand pressure without deformation or scratching.

hazing—clouding.

heavy centered pattern—spray pattern having most paint in center, less at edges.

hiding power—ability to obscure underlying surface.

★**high boiling solvent**—solvent with a high boiling point such as diacetone alcohol or cellosolve acetate.

high build—producing thick dry films per coat.

high flash naphtha—aromatic solvent having a high flash point, (min. 113°F, 45°C).

hold out—ability (or property) to prevent soaking into substrate.

holiday—pinhole; skip; discontinuity; void.

holiday detector—device for detection of pinholes or holidays. See spark testing.

hot spray—spraying material heated to reduce viscosity.

humidity—measure of moisture content; relative humidity is the ratio of the quantity of water vapor in the air compared to the greatest amount possible at the given temperature. Saturated air is said to have a humidity of 100 per cent.

hydraulic spraying—spraying by hydraulic pressure. (See airless spraying.)

I

incompatibility—inability to mix with or adhere to another material.

indictor (pH) paper—a vegetable dyed paper indicating relative acidity or basicity (alkalinity).

inert pigment—a non-reactive pigment.

inflammability—measure of ease of catching fire; ability to burn; use of the word flammability is preferred to inflammability due to the possibile misinterpretation of the prefix "in" as a negative.

inhibitive pigment—one which retards the corrosion process.

inorganic coatings—those employing inorganic binders or vehicles.

intermediate coat—middle coat; guide coat.

internal mix—a spray gun in which the fluid and air are combined before they leave the gun.

intumesce—to form a voluminous char on ignition; foaming or swelling when exposed to flame.

iron phosphate coating—conversion coating; chemical deposit.

isocyanate resins—urethane resins.

K

KB (Kauri-Butanol) Value—measure of solvent power.

ketones—flammable organic solvents; commonly used ketones are acetone; methyl ethyl ketone (MEK); and methyl isobutyl ketone (MIBK).

Krebs Units (K.U.)—arbitrary units of viscosity.

L

lacquers—coatings which dry by evaporation of the solvent.

laitance—milky white deposit on new concrete.

laminar scale—rust formation in heavy layers.

latex—rubber like; a common binder for emulsion (water) paints; there are natural and synthetic latexes.

leafing—orientation of pigment flakes in horizontal planes.

leveling—flowing out to films of uniform thickness; tendency of brush marks to disappear.

lifting—softening and raising of an undercoat by application of a top coat.

livering—formation of curds or gelling.

long oil varnish—varnish with a high ratio of oil to resin; a resin having a large quantity of oil cooked per 100 pounds of resin (25 gallons or more per 100 pounds of resin).

★**low boiling solvent**—solvent with a low boiling point such as acetone or methyl alcohol.

low pressure spraying—conventional air spraying.

M

MAC (maximum allowable concentration)—maximum concentration of solvent vapor in parts per million parts of air in which a worker may work eight consecutive hours without an air fed mask; the lower the MAC number, the more toxic the solvent.

maintenance painting—(1) repair painting; any painting after the initial paint job; in a broader sense it includes painting of items installed during maintenance; (2) all painting except that done solely for aesthetics.

mandrel test—a physical bending test for adhesion and flexibility.

masking—covering areas not to be painted.

mastic—a heavy bodied high build coating.

(MEK) methyl ethyl ketone—a strong flammable organic solvent.

(MIBK) methyl isobutyl ketone—a strong flammable organic solvent.

mil—one one-thousandth of an inch; .001"; 1/1000 in.

mildew—fungus, mold.

mildewcide—substance poisonous to mildew; prevents or retards growth of mildew.

mild steel—structural steel; SAE 1020.

mill scale—oxide layer formed on steel by hot rolling.

mineral spirits—aliphatic hydrocarbon solvent.

miscible—capable of mixing or blending uniformly.

misses—holidays; skips; voids.

mist-coat—thin tack coat; thin adhesive coat.

moisture and oil separator—trap on air compressor or in air lines.

mottling—speckling; a nonuniform paint color.

mud-cracking—irregular cracking of dried film, as in a dried mud puddle.

multicolor finishes—speckled finishes; paints containing flecks of colors different from the base color.

MVT (moisture vapor transmission)—moisture vapor transmission rate through a known membrane.

N

naphtha—flamable aliphatic hydrocarbon solvent.

near-white blast cleaning—see 4.4.2.4.

needle (spray gun)—fluid metering pin.

neoprene—a rubber-like film former based on the polymerization of chloroprene.

non-drying oil—one which will not oxidize in air.

non-ferrous—containing no iron.

non-flammable—incombustible, will not burn.

non-toxic—not poisonous.

non-volatile—solid; non-evaporating; the portion of a paint left after the solvent evaporates.

O

oil color—coloring (pigment or dye) dispersed in oil.

oil length—gallons of oil reacted with 100 pounds of resin.

oleoresinous—film former containing oil and resin.

opacity—hiding power.

orange peel—dimpled appearance of dry film; resembling an orange peel.

organic—containing carbon compounds.

organosol—film former containing resin plasticizer and solvent.

orifice—opening; hole.

overatomized—dispersed too finely by use of excessive atomizing air pressure.

overcoat—second coat; top coat.

overlap—portion (width) of fresh paint covered by next layer.

overspray—sprayed paint which did not hit target; waste.

P

PVAc—see polyvinyl acetate.

PVC—see polyvinyl chloride or pigment volume concentration.

paint—all coating materials used in painting.

paint failure—the loss of usefulness of the paint coating.

paint heater—device for lowering viscosity of paint by heating.

paint program—comprehensive painting plan.

paint project—single paint job.

paint system—the complete number and type of coats comprising a paint job. In a broader sense, surface preparation, pre-treatments, dry film thickness, and manner of application are included in the definition of a paint system.

painting—all operations required to use paints properly.

painting materials—all materials required to adequately paint a surface.

pass (spray)—motion of the spray gun in one direction only.

passive defense—blending of colors to make structures less conspicuous.

pattern length—length of spray pattern.

pattern width—width of a spray pattern at vertical center.

peeling—failure in which paint curls or otherwise strips from substrate.

perm—unit of permeance; grains of water vapor per hour per square foot per inch of mercury—water vapor pressure difference.

phenolic resins—particular group of film formers; resins made from phenols and aldehydes.

phosphatize—form a thin inert phosphate coating on surface usually by treatment with phosphoric acid or other phosphate compound.

phthalic resins—a particular group of film formers; alkyd resins.

pH value—measure of acidity or alkalinity; pH 7 is neutral; the pH values of acids are less than 7, and of alkalis (bases) greater than 7.

pickling—a dipping process for cleaning steel and other metals; the pickling agent is usually an acid.

pigment grind—dispersion of pigment in a liquid vehicle.

pigments—solid coloring agents.

pigment volume concentration (PVC)—percent by volume occupied by pigment in dry film.

pitting—formation of small, usually shallow depressions or cavities.

pin-holing—formation of small holes through the entire thickness of coating; see cratering.

plasticizer—a paint ingredient which imparts flexibility.

plastisol—film former containing resin and plasticizer with no solvents.

pock marks—pits; craters.

pole-gun—spray gun equipped with an extension tube.

polymer—a large molecule formed by polymerization.

polymerization—chemical reaction in which small molecules combine to form large molecules.

polyvinyl acetate (PVAc)—a synthetic resin used extensively in emulsion (water) paints; produced by the polymerization of vinyl acetate.

polyvinyl chloride (PVC)—a synthetic resin used in solvent type coatings produced by the polymerization of vinyl chloride.

porosity—degree of integrity or continuity.

pot-life—time interval after mixing of reactive components during which liquid material is usable with no difficulty.

pressure balance—in spray painting, relationship of pot pressure to atomizing air pressure.

pressure drop—loss in pressure due usually to length or diameter of line or hose.

pressure feed—fluid flow caused by application of air or hydraulic pressure to paint.

pressure feed paint tank (pressure pot)—fluid container in which fluid flow is caused by air pressure.

pretreatment—chemical alteration of the surface to make it suitable for painting.

preventive maintenance painting—period touch-up painting or application of full coats of paint before deterioration starts.

prime coat—first coat on a substrate.

primer—material used for prime coat; usually a rust-inhibitive coating when used over ferrous metals.

production rate (sq. ft./day)—measurement of surface area cleaned or coated in one working day by one man.

profile—surface contour of a blast-cleaned surface as viewed from the edge; cross section of the surface.

profile depth—average distance between top of peaks and bottom of valleys on the surface.

proprietary—available on open market under brand name.

protective life—interval of time during which a paint system protects substrate from deterioration.

pump ratio—multiplier of input pressure to indicate output pressure; ratio of air piston area to fluid piston area.

R

reaching (spray gun)—extending spray stroke too far.

rebound—paint spray bounce back. See bounce back.

recoat time—time interval needed between application of successive coats.

red label—flammable or explosive materials with flash points below 80°F. (26.7°C).

reducer—a material which lowers viscosity but is not necessarily a solvent for the particular film-former; thinner.

reflectance—degree of light reflection.

repainting—a complete painting operation including surface preparation.

repair of surfaces—all procedures required to return the surface to a satisfactory condition for painting.

resin—a material, natural or synthetic, contained in varnishes, lacquers and paints; the film former.

respirator—safety breathing mask.

rise—height.

roller coating—the act of painting with a roller; the material used for roller painting.

round pattern—circular spray pattern.

runs—curtains; sags.

rust—corroded iron; red iron-oxide deposited on metal; also other metal oxides formed by corrosion.

rust bloom—discoloration indicating the beginning of rusting.

S

safety valve—pressure release valve preset to be released when pressure exceeds a safe operating limit.

sandblast—blast cleaning using sand as an abrasive.

sandy finish—a surface condition having the appearance of sandpaper; may result from overspray.

saponify—convert to soap.

scale—rust occurring in thin layers.

scaler—a hand cleaning chisel.

scaling—process of removing scale.

seal coating—coating used to prevent excessive absorption of the first coat of paint by the substrate; a primer.

sealer—a low viscosity (thin) liquid sometimes applied on wood, plaster, gypsum board, or masonry.

seeding—formulation of small agglomerates.

separation—division into components or layers by natural causes.

settling—caking; sediment.

shade—degree of color in a tint.

shelf-life—maximum interval in which a material may be stored and still be in usable condition.

shop coat—coating applied in fabricating shop.

short oil varnish—a varnish prepared by cooking a relatively small quantity of oil with 100 pounds of resin, quick drying; brittle; less than 25 gallons of oil per 100 pounds of resin.

shot blasting—blast cleaning using steel shot as the abrasive.

shrinkage—decrease in volume on drying.

silicate paints—those employing silicates as binders; used primarily in inorganic zinc rich coating.

silicone resins—a particular group of film formers; used in water-repellent and high-temperature paints; organo-silicon polymers.

silking—a surface defect characterized by parallel hair-like striations in coated films.

skinning—formation of a solid membrane on top of a liquid.

skips—holidays; misses; uncoated area; voids.

slow drying—requiring 24 hours or longer before recoating is possible.

solids—non-volatile portion of paint.

solids by volume—percentage of total volume occupied by non-volatiles.

solubility—degree to which a substance may be dissolved.

solution—a liquid in which a substance is dissolved.

solvency—measure of ability to dissolve.

solvent—a liquid in which another substance may be dissolved.
solvent balance—ratio of amounts of different solvents in a mixture of solvents.
solvent pop—blistering caused by entrapped solvent.
solvent release—ability to permit solvent to evaporate.
solvent wash—cleaning with solvent.
spalling—the cracking, breaking or splintering of materials, usually due to heat or freezing.
spark testing—detection of holidays (flaws). Using a special spark testing tool. See holiday detector.
spark-proof tools—bronze beryllium tools.
spar varnish—a varnish for exterior surfaces.
specular gloss—mirror-like reflectance.
spray cap—front enclosure of spray gun equipped with atomizing air holes.
spray head—combination of needle, tip, and air cap.
spray pattern—configuration of spray with gun held steady.
spreading rate—area covered by a unit volume of coating frequently expressed as square feet per gallon.
SSPC—Steel Structures Painting Council
steam clean—a cleaning process using live steam.
streaks—a surface defect characterized by essentially parallel lines of different colors or shades.
stroke (spray)—a single pass with a spray gun in one direction.
styrene-butadiene resin—a copolymer of styrene and butadiene.
substrate—basic surface.
suction feed (sandblaster)—one in which the abrasive is syphoned to the nozzle.
suction feed (spray gun)—one in which the fluid is syphoned to the spray head.
surface conditioner—preparatory coating applied to chalked, painted masonry surface for bonding chalk to under surface.
surface preparation—all operations necessary to prepare a surface to receive a coating of paint.
surfacer—a paint used to smooth the surface before finish coats are applied.
sweating—condensing moisture on a surface.

T

tack—degree of stickiness.
tail line—short piece of blast hose smaller than the main hose to permit better maneuverability.
tails (airless spray)—finger-like spray pattern.
tank white—good hiding, self-cleaning white paint for exterior metal surfaces.
tapered pattern—elliptical shaped spray pattern; a spray pattern with converging lines.
tape test—a particular type of adhesion test.
test pattern—spray pattern used in adjusting spray gun.
thermoplastic—becomes mobile or softens under heat.
thermosetting—becomes rigid under heat and cannot be remelted.
thinners—volatile organic liquids for reducing viscosity; solvents.
thixotropic—a gel which liquifies with agitation but gels again on standing.
through dry—ability of film to show no loosening, detachment, or evidence of distortion when the thumb, placed on film with maximum arm pressure, is turned through 90° in plane of film.
tie coat—intermediate coat used to bond different types of paint coats.
tint—a color produced by the mixture of white paint or pigment in a predominating amount with a non-white colored paint or pigment.
tone down—the process of reducng visual prominence of an installation by the application of external coatings; blending of overall color scheme with the surrounding environment.
tooth—profile; mechanical anchorage; surface roughness.
top coating—finish coat.
touch-up painting—spot repair painting usually conducted after initial painting.
toxic—poisonous.
toxicity—degree of poisonousness or harmfulness.
transition primer (block or barrier coat)—coating compatible with primer and with a finish coat, though the latter is not compatible with the primer.
triggering—intermittent squeezing and releasing of trigger.
two-component gun—one having two separate fluid sources leading to spray head, for spraying a coating and a catalyst simultaneously.

U

underatomized—not dispersed or broken up fine enough
unit cost—cost per given area.
urethane resins—a particular group of film formers, i.e isocyanate resins.
useful life—the length of time a coating is expected to remain in service.

V

VM&P naphtha—varnish and paint manufacturers naphtha; a low power flammable hydrocarbon solvent.
vapor degreasing—a cleaning process utilizing condensing solvent as the cleaning agent.
vaporization—conversion from liquid or solid to a gaseous state.
varnish—liquid composition of oil, resin thinners and driers, which converts to a transparent or translucent solid film after application as a coating.
vehicle—liquid carrier; binder; anything dissolved in the liquid portion of a paint is a part of the vehicle.
vinyl coating—one in which the major portion of the binder is of a vinyl resin.
vinyl copolymer—resins produced by copolymerizing vinyl monomers such as vinyl acetate and vinyl chloride.
vinyl resins—synthetic resins made from vinyl compounds such as vinyl acetate.

viscosity—a measure of fluidity.

viscosity cup—a device for measuring viscosity.

volatiles—fluids which evaporate rapidly.

volatile content—those materials which evaporate; usually expressed as a percentage.

W

washing—erosion of a paint film after rapid chalking.

wash primer—a thin rust-inhibiting paint which provides improved adhesion to subsequent coats.

water blasting—blast cleaning using high velocity water.

weld spatter—beads of metal left adjoining a weld.

wet edge—fluid boundary.

wet film gauge—device for measuring wet film thickness.

wet film thickness—thickness of liquid film immediately after application.

wet spray—spraying so that surface is covered with paint that has not started to dry.

wetting oils—products used to promote adhesion of applied coatings when all mill scale and rust cannot be removed.

white blast—see 4.4.2.4. blast cleaning to white metal.

wire brush—a hand cleaning tool comprised of bundles of wires; also the act of cleaning a surface with a wire brush, including power brushes.

wrinkling—a surface defect resembling the skin of a prune.

wrist action (spray gun)—swiveling of wrist without arcing forearm.

Y

yellowing—development of yellow color or cast, in whites, on aging.

Z

zinc phosphate coating—treatment used on steel to improve adhesion of coatings.

zinc silicate—inorganic zinc coating.

zinc yellow—commercial zinc chromate pigment.

ANSWER SHEET

TEST NO. ______________ PART ________ TITLE OF POSITION ______________________________
(AS GIVEN IN EXAMINATION ANNOUNCEMENT - INCLUDE OPTION, IF ANY)

PLACE OF EXAMINATION ______________________________ ______________________ DATE ______________
(CITY OR TOWN) (STATE)

RATING

USE THE SPECIAL PENCIL. MAKE GLOSSY BLACK MARKS.

1 A B C D E	26 A B C D E	51 A B C D E	76 A B C D E	101 A B C D E
2 A B C D E	27 A B C D E	52 A B C D E	77 A B C D E	102 A B C D E
3 A B C D E	28 A B C D E	53 A B C D E	78 A B C D E	103 A B C D E
4 A B C D E	29 A B C D E	54 A B C D E	79 A B C D E	104 A B C D E
5 A B C D E	30 A B C D E	55 A B C D E	80 A B C D E	105 A B C D E
6 A B C D E	31 A B C D E	56 A B C D E	81 A B C D E	106 A B C D E
7 A B C D E	32 A B C D E	57 A B C D E	82 A B C D E	107 A B C D E
8 A B C D E	33 A B C D E	58 A B C D E	83 A B C D E	108 A B C D E
9 A B C D E	34 A B C D E	59 A B C D E	84 A B C D E	109 A B C D E
10 A B C D E	35 A B C D E	60 A B C D E	85 A B C D E	110 A B C D E

Make only ONE mark for each answer. Additional and stray marks may be counted as mistakes. In making corrections, erase errors COMPLETELY.

11 A B C D E	36 A B C D E	61 A B C D E	86 A B C D E	111 A B C D E
12 A B C D E	37 A B C D E	62 A B C D E	87 A B C D E	112 A B C D E
13 A B C D E	38 A B C D E	63 A B C D E	88 A B C D E	113 A B C D E
14 A B C D E	39 A B C D E	64 A B C D E	89 A B C D E	114 A B C D E
15 A B C D E	40 A B C D E	65 A B C D E	90 A B C D E	115 A B C D E
16 A B C D E	41 A B C D E	66 A B C D E	91 A B C D E	116 A B C D E
17 A B C D E	42 A B C D E	67 A B C D E	92 A B C D E	117 A B C D E
18 A B C D E	43 A B C D E	68 A B C D E	93 A B C D E	118 A B C D E
19 A B C D E	44 A B C D E	69 A B C D E	94 A B C D E	119 A B C D E
20 A B C D E	45 A B C D E	70 A B C D E	95 A B C D E	120 A B C D E
21 A B C D E	46 A B C D E	71 A B C D E	96 A B C D E	121 A B C D E
22 A B C D E	47 A B C D E	72 A B C D E	97 A B C D E	122 A B C D E
23 A B C D E	48 A B C D E	73 A B C D E	98 A B C D E	123 A B C D E
24 A B C D E	49 A B C D E	74 A B C D E	99 A B C D E	124 A B C D E
25 A B C D E	50 A B C D E	75 A B C D E	100 A B C D E	125 A B C D E

ANSWER SHEET

TEST NO. ______________ PART ______ TITLE OF POSITION ______________________________
(AS GIVEN IN EXAMINATION ANNOUNCEMENT - INCLUDE OPTION, IF ANY)

PLACE OF EXAMINATION ______________________________ ______________ DATE ______________
(CITY OR TOWN) (STATE)

RATING

USE THE SPECIAL PENCIL. MAKE GLOSSY BLACK MARKS.

	A	B	C	D	E		A	B	C	D	E		A	B	C	D	E		A	B	C	D	E		A	B	C	D	E
1						26						51						76						101					
2						27						52						77						102					
3						28						53						78						103					
4						29						54						79						104					
5						30						55						80						105					
6						31						56						81						106					
7						32						57						82						107					
8						33						58						83						108					
9						34						59						84						109					
10						35						60						85						110					

Make only ONE mark for each answer. Additional and stray marks may be counted as mistakes. In making corrections, erase errors COMPLETELY.

	A	B	C	D	E		A	B	C	D	E		A	B	C	D	E		A	B	C	D	E		A	B	C	D	E
11						36						61						86						111					
12						37						62						87						112					
13						38						63						88						113					
14						39						64						89						114					
15						40						65						90						115					
16						41						66						91						116					
17						42						67						92						117					
18						43						68						93						118					
19						44						69						94						119					
20						45						70						95						120					
21						46						71						96						121					
22						47						72						97						122					
23						48						73						98						123					
24						49						74						99						124					
25						50						75						100						125					